KEROSENE HEATERS

BY DAN RAMSEY

TAB **TAB BOOKS Inc.**
BLUE RIDGE SUMMIT, PA. 17214

FIRST EDITION

FIRST PRINTING

Copyright © 1983 by TAB BOOKS Inc.

Printed in the United States of America

Reproduction or publication of the content in any manner, without express
permission of the publisher, is prohibited. No liability is assumed with respect to
the use of the information herein.

Library of Congress Cataloging in Publication Data

Ramsey, Dan.
Kerosene heaters.

Includes index.
1. Kerosene heaters. I. Title.
TH7450.5.R35   1983   697'.24   83-4919
ISBN 0-8306-0198-8
ISBN 0-8306-0598-3 (pbk.)

Cover photographs courtesy of Kero-sun, Inc.

Contents

Acknowledgments

I GRATEFULLY ACKNOWLEDGE THE HELP OF MANY sources in preparing this book. In random order, they include: Jerry Bjork, Public Affairs Specialist, U.S. Consumer Safety Commission; Jack E. Shaffer, Standards Engineer, Standards Department, Underwriters Laboratories, Inc.; Roger D. Mitchell, Senior Vice President of Marketing, Fanco USA; Trisha Funk, Supervisor of Public Relations and Edward F. Miller, Director of Public Relations, Kero-Sun Inc.; George G. Noll, American Petroleum Institute; S. Mitch Meyers, National Consumer Relations Manager, Robeson Industries, Inc.; Mark M. DiPierro, Fire Protection Engineer, Engineering and Safety Service, American Insurance Association; GLO International Corporation; Raymond B. Penkola, Assistant Product Manager, Teknika Electronics Corporation; Walter S. Harrah, Vice President of Public Affairs, National Kerosene Heater Association; Elsie Deatherage, Consumer Education Representative, Public Utility District of Clark County (Washington); Washington State University, Cooperative Extension Service; Cornell University Cooperative Extension Program; U.S. Department of Agriculture's Forest Service and Agricultural Research Service. Special acknowledgment goes to the management of Dunn-Rowan, Inc. for their support.

Most important, I gratefully acknowledge the gifts of love, concern, and friendship by my first editor, Judy. This book and I are dedicated to you.

Introduction

MORE THAN 3 MILLION AMERICANS WILL BE TURNING on kerosene heaters in their homes this winter. The number is expected to *triple* in the next few years. Yet some experts say that portable kerosene heaters are unsafe.

What are these heaters, and why are they so controversial? Kerosene heaters, mostly imported from Japan where kerosene heat has safely warmed homes since the 1950s, are supplemental space heaters producing 7,000 to 18,000 British thermal units (Btu) of heat per hour. They initially cost more than other space heaters—from $85 to $300 each—but are cheaper to operate, often costing half as much as gas or electricity. They are 95 percent efficient and don't require a flue or wiring.

This book will take you step by step from a basic understanding of why and how the portable kerosene heater works, through selection and use, increasing efficiency, improving safety, maintenance and repair, selecting accessories, and installing larger kerosene heaters. A glossary will help you through the vocabulary of these popular appliances. Everything you need to know is here!

Special Notice

Before purchasing any portable kerosene heater for use in a home, rental property, or business location, check with local fire officials and your insurance agent on the legality, requirements, and use of such heaters. Numerous state and local regulations ban or limit the use of portable kerosene heaters.

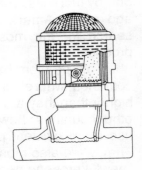

Chapter 1

The Kerosene Heater

THE REVIVAL OF WOOD STOVES RECEIVED MUCH PUBlicity during the past few winters as homeowners searched for alternative fuels. More recently, millions of homes and apartments are being given auxiliary heat with another old-time fuel—kerosene.

Portable kerosene room heaters can be reliable and cost-efficient alternatives for home heating if they contain appropriate safety features. Modern kerosene heaters have been developed in recent years with features that set them apart from primitive designs common a few decades ago. Kerosene heaters have become "hot items."

THE NEW KEROSENE HEATER

Space heaters have always been useful for providing warmth in partially heated or unheated areas of the house. In recent years they have helped to reduce home heating bills that seem to rise with each heating season. A space heater is designed so that you can lower the central heating system thermostat and still keep a room comfortably warm.

Space heaters have taken many forms—from fireplaces to charcoal burners to electric radiant heaters. Kerosene heaters provided warmth on farms and in homes and shops across the nation. Many people recall them as smelly, smoky devices that users were happy to set aside for introduction of low-cost natural gas, electricity, and heating oil.

Technology has caught up with the kerosene heater, which is related to the old kerosene heater only by fuel lines. The OPEC oil embargo of a decade ago made Americans consider alternatives and search for the most efficient users of petroleum products.

This problem had already been faced in many countries where oil prices have long been much higher than they are in the United States. Japan and other countries show Americans how to produce and use more efficient petroleum-burning automobiles. They were also pioneers in the development of the "new" kerosene heater. Today, unvented kerosene heaters are the major source of residential heating in Japan. There are 1.4 such heaters per household in that country—38 million units. Europe, too, leads America in the production and use of kerosene heaters.

HEATER SAFETY

The growth of kerosene heaters has been restrained by concerns for safety. Reports pro and con have circulated on the safety hazards of unvented kerosene heaters due to oxygen consumption, carbon monoxide production, and potential fire hazards. I will address these important safety considerations throughout this book. In addition, Chapter 6 will outline how to improve kerosene heater safety.

Kerosene heaters can be used for many years without problems if the consumer understands the concepts, uses, and limitations of these heaters. This book attempts to educate you the consumer, on the safe and efficient selection, purchase, use and maintenance of kerosene heaters.

HOW THE HEATER WORKS

While types and brands of kerosene heaters may be confusing in number, all work on the same basic principle. They all involve four easy-to-understand systems.

Fuel System and Wick

See Fig. 1-1. Kerosene in the fuel tank or *sump* is absorbed by the wick skirt, which is made of cotton

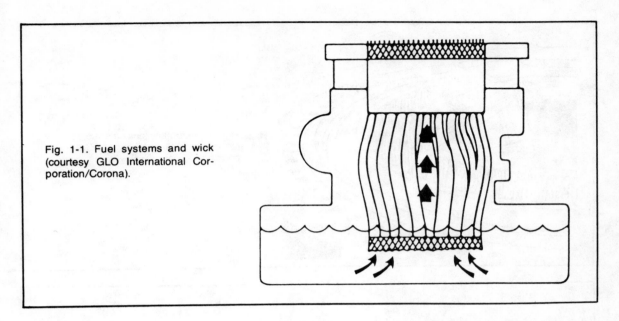

Fig. 1-1. Fuel systems and wick (courtesy GLO International Corporation/Corona).

fibers. The fuel is then pulled to the fiberglass portion of the wick by *capillary action*.

Automatic Ignition System

At the wick's fiberglass top, the kerosene is ignited for primary combustion by a battery-powered ignition plug (Fig. 1-2). This plug is briefly switched on and put in contact with the wick when the ignition system's operating lever is pushed. The plug heats to about 1850° F. (1000° C.) and instantly vaporizes and ignites the kerosene at the wick's top. The kerosene continues to vaporize and burn in a process called *primary combustion*.

Burner Assembly

The burner assembly (Fig. 1-3) does two jobs. First, it controls the rate and volume of fresh air drawn through the base of the heater. Second, it reburns the kerosene vapors produced by the primary combustion process at the wick. This happens in the burner's baffled chambers or chimneys and is called *secondary combustion*. This reburning process makes the kerosene heater highly efficient and eliminates the need for a vent.

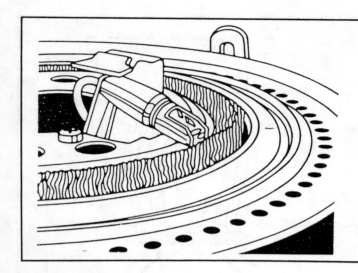

Fig. 1-2. Automatic ignition systems (courtesy GLO International Corporation/Corona).

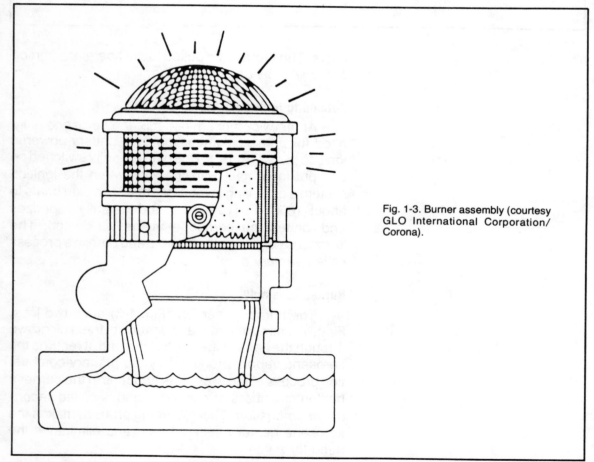

Fig. 1-3. Burner assembly (courtesy GLO International Corporation/Corona).

Fig. 1-4. Wick adjustment knob (courtesy GLO International Corporation/Corona).

Wick Adjustment and Automatic Shutoff System

Turning the wick adjusting knob (Fig. 1-4) causes the wick adjusting mechanism to raise or lower the entire wick for precise burning exposure at the base of the burner assembly. When the wick is raised higher, resulting heat output will be proportionately higher. In Fig. 1-5 a ratchet wheel attached to the spring-wound wick shaft of the wick adjusting mechanism holds the wick at the desired height during normal operation. When bumped, jarred excessively, or overturned, the heater's automatic shutoff system pendulum weight triggers the release of the ratchet lever with the ratchet wheel. This allows the wick to instantly retract from the burning position. Some kerosene heaters use similar mechanisms to do the same job: turn the burner off if the unit is knocked over.

TYPES OF HEATERS   There are two types of kerosene heaters marketed today: the *convection* heater (Figs. 1-6 and 1-7) and the *radiant* heater (Figs. 1-8 and 1-9). Most convec-

Fig. 1-5. Automatic shutoff system (courtesy GLO International Corporation/Corona).

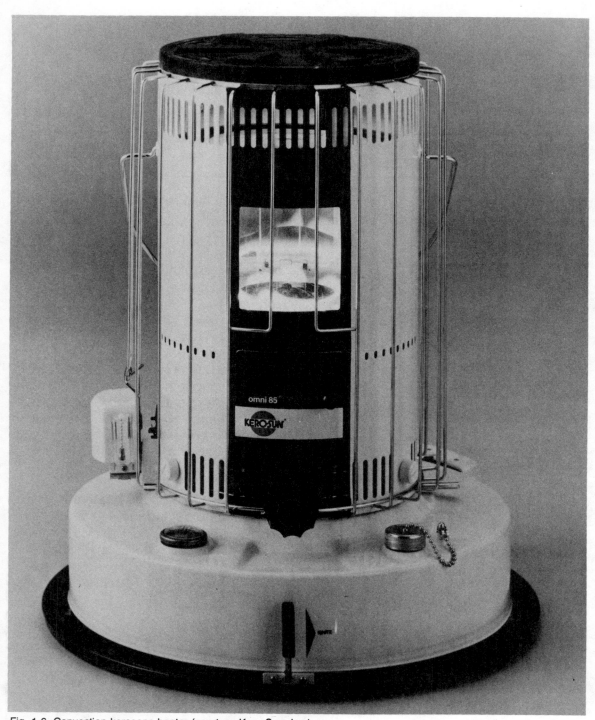

Fig. 1-6. Convection kerosene heater (courtesy Kero-Sun, Inc.).

Fig. 1-7. Convection kerosene heater (courtesy Teknika Electronics Corporation).

Fig. 1-8. Radiant kerosene heater (courtesy Kero-Sun, Inc.).

tion heaters are round, and most radiant heaters are rectangular.

Convection heaters are designed to move warmed air up and away from the heater (Fig. 1-10). This forces cool room air down and back toward the base of the heater where it is warmed. Convection heaters are normally placed near the center of large, often hard to heat areas. Most have these three similarities:

- They are circular in shape, with a grille around most or all of the burner area.
- The base of the heater includes an integral fuel tank.
- A reflector is not used behind the burner as in radiant model heaters.

Fig. 1-9. Radiant kerosene heater (courtesy Teknika Electronics Corporation).

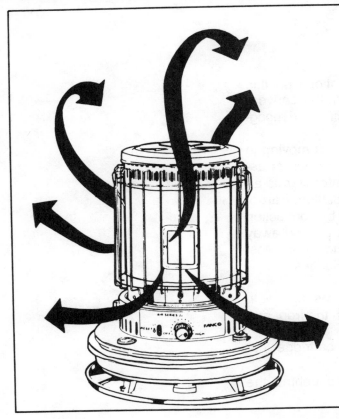

Fig. 1-10. Flow of heat from convection heater (courtesy Fanco USA).

Figure 1-11 illustrates the typical kerosene convection heater. Here are the basic components:

A   Inspection window.
B   Burner assembly flame spreader.
C   Grille guard.
D   Burner assembly.
E   Ignition glow plug.
F   Pendulum and automatic shutoff system.
G   Automatic shutoff system reset lever.
H   Fuel tank.
I   Wick adjustment knob.
J   Automatic ignition system operating lever.
K   Top plate.
L   Heater cabinet.
M   Carrying handle.
N   Wick top.
O   Wick adjusting mechanism.
P   Wick skirt.
Q   Heater base.

There are numerous minor variations on this basic model, but becoming familiar with this convective heater will give you an understanding and recognition of nearly all of them.

Radiant heater models provide heat moving in straight waves directly from the heater's burner assembly and reflector. This type of heater warms air and items in front of the heater by radiation. Part of the total heat output is also circulated by convection to the surrounding areas as it moves up and away from louvers in the top of the cabinet (Fig. 1-12). Most radiant heaters have these three similarities:

• They are rectangular or boxy in shape, with most all heat moving away from the front and top of the heater.
• The fuel tank is a removable cartridge that allows outdoor filling.
• A polished reflector is located behind the burner assembly.

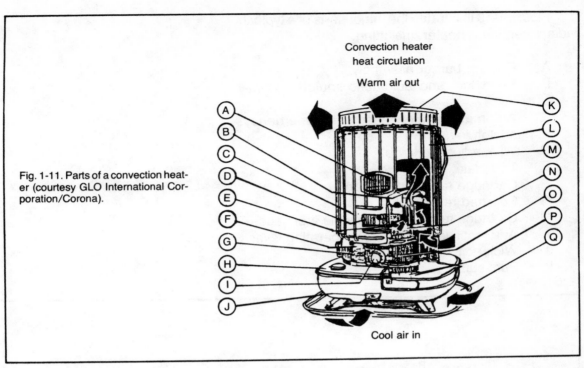

Convection heater
heat circulation

Warm air out

A
B
C
D
E
F
G
H
I
J

K
L
M
N
O
P
Q

Fig. 1-11. Parts of a convection heater (courtesy GLO International Corporation/Corona).

Cool air in

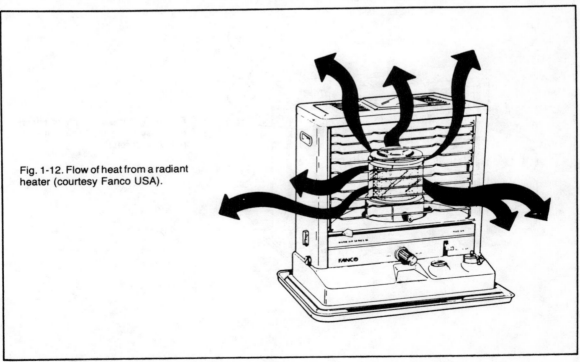

Fig. 1-12. Flow of heat from a radiant heater (courtesy Fanco USA).

Figure 1-13 illustrates the major parts of a typical radiant kerosene heater, including:

A   Radiant burner assembly.
B   Pendulum and automatic shutoff assembly.
C   Automatic shutoff assembly emergency shutoff lever.
D   Wick adjusting knob.
E   Top plate.
F   Cartridge tank access door.
G   Polished reflector.
H   Cartridge fuel tank with visible fuel gauge.
I   Automatic ignition system operating lever.
J   Wick adjusting mechanism.
K   Fuel sump.

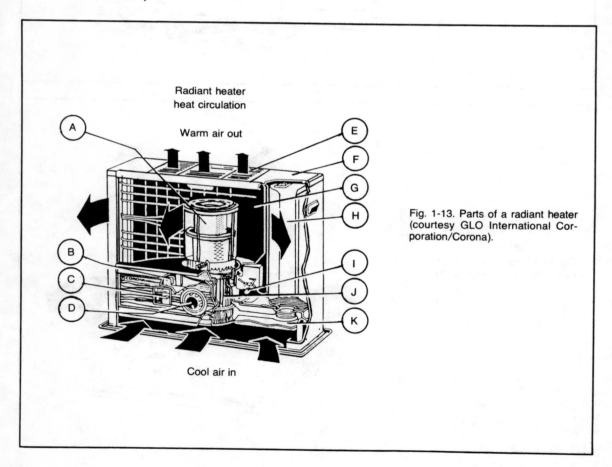

Fig. 1-13. Parts of a radiant heater (courtesy GLO International Corporation/Corona).

WICK   Another important difference between most convective and radiant kerosene heaters is how the wick is fed. As noted earlier, fuel around the skirt of the wick is absorbed by cotton strands to be carried up the wick. Above the saturated cotton wick skirt are thousands of very tiny fiberglass strands. See Fig. 1-14.

These small diameter strands are woven so tightly together that the fuel is pulled upward in each tiny space formed between the strands by capillary action. It takes slightly less than 30 minutes for the fuel to move from the wick skirt to the top of the fiberglass wick where combustion occurs.

Wicks in the radiant and convective heaters are the same. The process of moving the fuel is different in each.

Figure 1-15 illustrates how fuel flows in convection heaters. First, the kerosene fuel is siphon pumped into the tank in the heater base. Gravity keeps the fuel at the lowest level in the tank. Putting

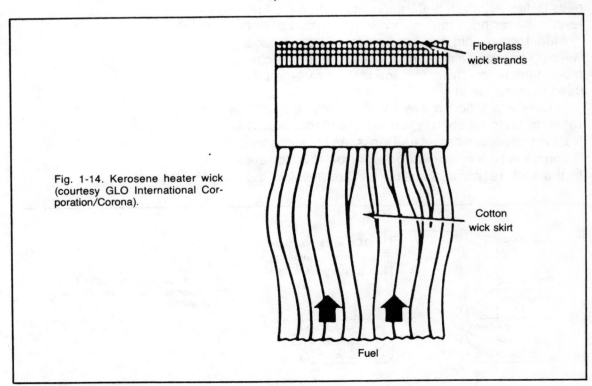

Fig. 1-14. Kerosene heater wick (courtesy GLO International Corporation/Corona).

Fiberglass wick strands

Cotton wick skirt

Fuel

fuel in the tank raises the level on the wick skirt and reduces the distance fuel must travel before it reaches the wick's top surface.

Capillary action draws fuel up the wick fibers to the wick top for combustion. As the fuel drops lower in the tank, it must travel further up the wick. Capillary action slows. Combustion output diminishes as low fuel level reduces fuel flow. If the fuel level falls below the bottom of the wick skirt, the fuel flow up the wick stops, the wick burns dry, and combustion stops.

The fuel flow in radiant kerosene heaters is slightly different (Fig. 1-16). Kerosene fuel is siphon pumped into the cartridge tank. The cap or fuel feeding fitting is replaced on the filled cartridge tank to keep the fuel from spilling out. The inverted cartridge tank is placed in the heater with the needle valve in the fuel feeding fitting opened by the fuel receiver pin.

Gravity causes fuel to flow or gurgle out of the cartridge into the cartridge tank reservoir, then down the connector tube to the wick reservoir. Fuel level rises or fills evenly in both reservoirs. When the fuel level in the cartridge tank reservoir rises to the base of the fuel receiver pin, the fuel flow from the cartridge will stop because air can no longer go from the cartridge tank reservoir up into the tank to replace fuel used in combustion.

Capillary action draws the fuel from the wick reservoir up to the fiberglass burning surface, exactly like the convection heater fuel tanks. As the kerosene is consumed during combustion, the fuel level lowers in the wick reservoir, causing the fuel level in the

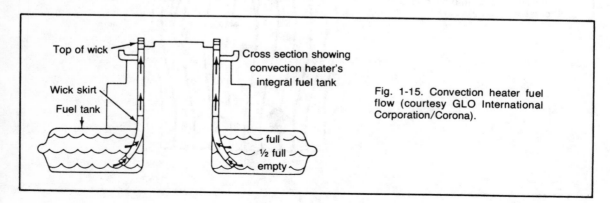

Fig. 1-15. Convection heater fuel flow (courtesy GLO International Corporation/Corona).

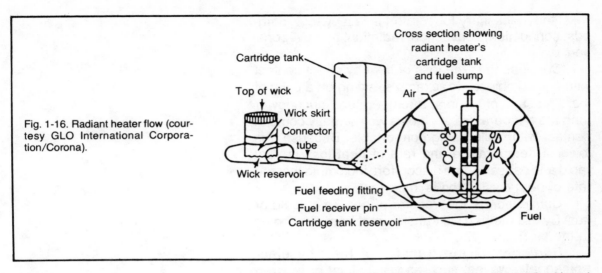

Fig. 1-16. Radiant heater flow (courtesy GLO International Corporation/Corona).

Cross section showing radiant heater's cartridge tank and fuel sump

Cartridge tank

Top of wick

Wick skirt

Connector tube

Wick reservoir

Air

Fuel feeding fitting

Fuel receiver pin

Cartridge tank reservoir

Fuel

cartridge tank reservoir to also lower. As the fuel level lowers in the cartridge tank reservoir, an air space is created between the level of the fuel and the fuel feeding fitting (tank cap). The air space allows air to enter the cartridge tank and displace any fuel contained within the tank.

As fuel drains down from the cartridge tank and raises the fuel level in the cartridge tank reservoir, the volume of the air space between the fuel level and fuel feeding fitting (cap) decreases. Fuel cannot flow from the cartridge tank into the cartridge tank reservoir unless there is an air space existing between the fuel level and the fuel feeding fitting.

RESIDENTIAL HEATING

A residential heating system should provide a comfortable atmosphere for the occupants. Let's examine the factors associated with heat that have a bearing on comfort.

The body must lose heat to be comfortable. Everything we do and all the body processes produce heat. Unless the heat generated can be dissipated at the same rate it is produced, body temperature will either rise or fall. If body temperature changes just a degree or two, acute discomfort is felt. Normally body temperature is 98.6° F.; however, the surface of exposed skin and clothing under comfortable conditions averages only about 82° F.

Heat necessary for comfort is lost by four methods: conduction, convection, radiation, and evaporation.

Conduction is the loss of heat by direct contact with a solid object—for example, sitting on a cake of ice. The cake of ice is at a temperature much lower than the temperature of objects we normally come in contact with while being comfortable. Because it is lower in temperature, the rate of heat loss is very rapid and causes some discomfort, indicating that the rate of heat dissipation is important.

Convection is the transfer of heat to a liquid or gas. By this method we transfer heat from our bodies to the air (Fig. 1-17).

Radiation is the transfer of heat by electromagnetic waves, such as visible light or infrared rays. The infrared rays are most important in heating. They travel through space and heat the objects they strike. Heat rays flow from a warm object to a cooler object. We receive heat rays from the sun or from any warmer object. We radiate heat to the nearby objects that are cooler than the surface of our skin or clothing, such as windows or walls (Fig. 1-18).

Evaporation is the changing of water from a liquid state to a vapor. Whenever this change of state takes place, heat is required. When perspiration evaporates from our skin, it takes heat from the skin's surface and thereby cools us. By respiration, when the air we breathe out carries more water vapor than

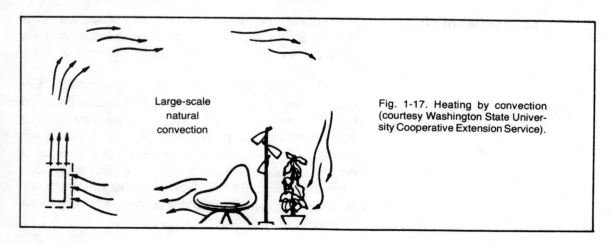

Large-scale natural convection

Fig. 1-17. Heating by convection (courtesy Washington State University Cooperative Extension Service).

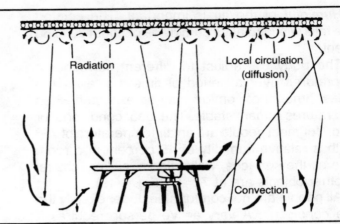

Fig. 1-18. Air circulation in a heated room (courtesy Washington State University Cooperative Extension Service).

Radiation

Local circulation (diffusion)

Convection

the air we breathe in, we are dissipating heat. The vapor added to the exhaled air takes heat from our lungs, which is in turn supplied from other parts of our body by blood circulation.

It takes 1,000 times as much heat to change water from liquid to vapor as it does to raise the temperature of liquid water 1° F. This process is also reversible. One thousand times as much heat is given up when vapor changes to water as when the temperature of water is lowered 1° F. This is a factor in causing condensation on cold surfaces.

The amount of heat that must be lost for comfort is not too important but may be of interest. An average adult sitting down loses about 400 Btu per hour. A Btu is the amount of heat required to raise 1 pound of water 1° F. Generally a person is comfortable when about one-fifth of the heat loss is dissipated by evaporation, about two-fifths by convection, and about two-fifths by radiation, with some being lost by conduction.

BODY HEAT LOSS FACTORS

Temperature is the main factor affecting body heat loss. This includes the temperature of the air and the temperature of all surfaces immediately surrounding a person.

The movement of air is important. Sitting in a draft is uncomfortable. Likewise, sitting in a room where the air is so stagnant that odors tend to build up is also somewhat distracting.

17

Humidity of the air is important. The comfort range falls within relative humidities of about 30 to 60 percent.

The body will adjust to different surrounding temperatures over a period of time; however, the greatest amount of comfort may be expected when temperatures remain stable. The ideal conditions for comfort for most people are an air temperature of 72° F. with a relative humidity of 45 percent in a room where all the surfaces, walls, floors, ceiling, chairs, and other objects are at 72° F. (Fig. 1-19).

All people are not comfortable under exactly the same conditions. A very active person requires a lower temperature of air and surroundings than an inactive person. A temperature of 72° F. is more suitable for an inactive person.

If the air temperature and surrounding temperatures are not all 72° F., and they never are, then for conditions of comfort we should attempt to get them as near 72° F. as possible. Again there will be a difference between active and inactive persons. The active person may feel more comfortable when a greater portion of his heat loss is dissipated by con-

Fig. 1-19. The comfortable atmosphere (courtesy Washington State University Cooperative Extension Service).

vection. A person sitting down and inactive may feel more comfortable if a slightly higher percentage of the heat dissipation is by radiation.

This explains why air movement that we might consider as a draft if we were sitting down resting would be very welcome if we were actively performing some task. If surrounding surfaces are cool, the temperature of the air must be warm to give us any degree of comfort. In situations where the surface temperatures are warm, the air temperature can be lower and we will still be rather comfortable.

We are most comfortable when all the surface temperatures around us are the same. If surface temperatures on one side are cold, such as a large picture window, and the surface temperatures on the other side of us are warm, we may be uncomfortable, even though the average is 72° F.

COMFORT AND HEATING SYSTEMS

There are numerous factors contributing to comfort that may differ from one system to another. They include air treatment, humidification, ventilation, surface temperatures, cleanliness and noise level.

Air is a most important commodity when considering heating systems. Its temperature is most important. Our comfort is quite easily affected by temperature. When we are at rest, a temperature difference of just 1 or 2 degrees is noticeable, so temperature fluctuation is important. Then there is the temperature difference from floor to ceiling. When the temperature varies from floor to ceiling as much as 10 to 15 degrees as it does in some heating systems, there is only one level in a room that would be comfortable. Above that it is too hot and below it is too cold. We want a small temperature difference from floor to ceiling.

Temperature response is important to comfort. If we turn the thermostat to 72° F. after it has been at 65° F. overnight, will the system bring up the general heat level quickly or does it take a long time? A quick response is desirable.

In dry areas and in areas where temperatures fall below zero, it is necessary as far as comfort is concerned to add humidity to the air. This is not important in humid areas. Perhaps the opposite is true; there is generally more humidity in the air than is desirable. There is, however, no simple method of reducing the relative humidity with a heating system. It can be done with a cooling system. A dehumidifier is an additional piece of equipment that may be generally desirable in humid areas of the country.

We want fresh air to supply our oxygen needs and to remove impurities and odors. Ventilation is also important in areas of high relative humidity. Moving air reduces the condensation of moisture on windows and walls and is a factor in minimizing mold that might otherwise accumulate in stagnant areas.

As mentioned earlier, surface temperatures go hand in hand with air temperature in providing comfortable conditions. We are therefore interested in evaluating the effect of any heating system on the surface temperature of the floor, ceiling, walls, and windows.

The temperature of the heat medium is a factor to evaluate in judging between heating systems. *Heat medium* is the temperature of the air as it is discharged into the room or the temperature of a radiation surface used to heat the room. The nearer this heat medium temperature is to the air temperature and the surface temperatures within the room, the better it will be for comfort. A small radiator can furnish enough heat to make a room comfortable. Because one part of the room, the radiating surface, is much hotter than all the other surfaces and the air temperature, it is less comfortable. If an entire wall, floor, or ceiling is the heat medium, it does not need to be nearly as warm to furnish the same amount of heat, simply because of its larger area.

CLEANLINESS AND NOISE

Most heating systems when operating properly do not produce dirt that enters into the room or circulating system. Soot production and odors are more related to the fuels than the type of heating system. The main problem with heating systems, as far as dirt is concerned, is dust circulation. The circulating air picks up the dust in the house, circulates it, and redeposits it on drapes, cold walls, and the like.

Another factor is whether or not the system will remove dust from the circulating air. Filters, such as those in ordinary forced air furnaces, remove large particles of dust. The small particles go on through and settle on furniture, drapes, and cold surfaces.

It is difficult to make any general statements concerning noise level because practically all systems are noisy under certain circumstances. The burning of oxygen and the expansion and contraction of heated metal in kerosene heaters produces some noise, but kerosene heaters are quieter than many systems.

COMPARING HEATERS

Portable kerosene heaters compare favorably with most types of heating systems, especially those *supplementary systems* that augment the home's main heating equipment. They treat the air by heating it with conduction or radiation, depending on the type of kerosene heater used. Humidity is maintained as a gallon of kerosene, when burned, emits a gallon of water into the air. Proper ventilation needed for kerosene heaters induces air movement into a room, reducing condensation and improving the room's comfort. Surface temperatures on a kerosene heater can be high, but adequate air movement can minimize this problem. If operated correctly, kerosene heaters are relatively clean and odor-free. Their noise output is minimal.

Portable electric heaters, by comparison, do not add humidity to the air. They produce only moderate circulation and ventilation and are very quiet to operate.

Chapter 2

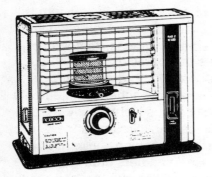

Selecting a Kerosene Heater

BUYING A KEROSENE HEATER IS SIMPLE. YOU JUST walk into the store, pick up a kerosene heater, and go pay for it.

Selecting a kerosene heater is not as simple. You must consider needs, finances, product claims, and efficiency. In this chapter you will learn how to select the safest and most appropriate kerosene heater based on knowledge and research. It takes a little more time than buying a kerosene heater, but the effort will build an awareness of the benefits and limitations of your heater that can add to its value.

Before considering the purchase of a portable kerosene heater, determine whether the device is legal for your intended use. The only reliable source of this information is your local fire official. In New York state portable kerosene heaters must be tested, inspected, and approved by a recognized independent agency such as Underwriters Laboratories. In some locations, heaters may be sold but not used. In other places, kerosene heaters are now allowed in multiple dwellings.

Even if your heater is legal in your area, contact your fire insurance agent before purchasing it. Unauthorized use may invalidate your policy.

Many states and cities require that all portable kerosene heaters be approved by the generally ac-

THE FIRST STEP

cepted Underwriters Laboratories Standard "UL 647." Listing guarantees that the heater meets standards for fuel capacity (less than 2 gallons), carbon monoxide output (less than .08 percent "worse case"), tip-over angle (less than 33 degrees from vertical), automatic shutoff upon tip-over, and proper labeling and operating instructions.

If you are considering buying or already own a portable kerosene heater, your first source of detailed information should be the owner's manual. Advertising copy normally doesn't offer the factual information needed to make the best decision. As you shop, have the dealer show you how to maintain the heater—especially cleaning, changing the wick, and fueling the device. Ask him to let you see the owner's manuals from numerous models for comparison.

COMPARING ELECTRIC AND KEROSENE HEATERS

Table 2-1 compares portable kerosene heaters to electric space heaters, which are perhaps the most well-known and widely available alternatives. This chart is for comparison use only.

Table 2-1. Comparison Chart for Portable Heaters (courtesy Cornell University Cooperative Extension).

Characteristics	Kerosene	Electric
Maximum output (approximate)	7000-20,000 Btu/hr. 2050-5850 watts	1700 Btu/hr. and up ιp 500 watts and up
Minimum output (approximate)	70 percent of maximum	10 percent of maximum (via on-off cycling)
UL listing available	Yes	Yes
Radiant models	Yes	Yes
Convection models	Yes	Yes
Required maintenance	Cleaning, fueling Wick replacement	Cleaning (dust)
Adds moisture to air	Yes	No
Odor	Possible smell of kerosene upon start-up/shutdown	Possible burning dust
Outdoor use	Yes	Generally no
Emergency use (power outage)	Yes	No
Requires ventilation	Yes	No
Burns	Possible	Possible
Shock	No	Possible
Asphyxiation	Possible	No
Other:	May affect house plants	Must be plugged into wall outlets

Voltage	× Amperage	= Wattage	= (× 3.412 watts) = Btu/hr.
115	15	1725	5886
115	20	2300	7848
230	20	4600	15695 (not portable)

Table 2-2. Potential Heat Output from Household Circuits (courtesy Cornell University Cooperative Extension).

An important difference is the heat output available. Electric heaters are limited by the amperage and voltage of available household circuits. Table 2-2 shows potential heat output from various household circuits. The typical portable heater is smaller—for example, a 1500-watt heater for a 15-amp, 115-volt circuit. Note that the heater controls allow you to vary output down to a small percentage of maximum. It is sometimes possible to use more than one heater in a room, but it is generally best to permanently wire a large heater.

Portable kerosene heaters are not recommended for use in small rooms. This is not only due to potential safety hazards, but is simply an indication of the large quantity of heat produced by these devices. Fuel-burning heaters work efficiently at one firing rate; they obtain the maximum heat from the fuel and produce minimum pollution emissions at this rate. Standard vented furnaces vary their output to satisfy the thermostat by burning for an interval, shutting off, then starting again when heat is needed. Because wick-fed devices such as portable kerosene heaters cannot shut off and restart automatically, their output is fairly constant. You can turn the wick down somewhat to obtain less heat, but this will decrease burning efficiency and increase objectionable emissions. The lower limit of output is about 70 percent of maximum. You can calculate it from the manufacturer's specifications for estimated burning time for one tank of fuel: divide the smallest number of hours (maximum output) by the largest number (minimum output). Remember that if the heater smokes or smells, you may have turned the wick too low.

The quality of kerosene varies depending upon refining practices and transport and storage methods.

FUEL

Kerosene deteriorates with prolonged storage and is easily degraded by water, particulates, mixtures of other fuels, or by high temperatures. The American National Standard for kerosene (ASTM D 3699-1978) specifies two grades of kerosene: 1-K (also called K-1) for nonvented units and 2-K (or K-2) for vented burners. The specifications differs only in maximum allowable sulfur content of purchased fuel. The required fuel tank warning reads:

CAUTION: Improper fuel may cause pollution and sooting of the burner. Use only low-sulfur, water-clear kerosene.

Any other fuel, such as No. 1 fuel oil, will decrease wick life and heater performance and increase maintenance requirements.

To monitor the kerosene, do the following:

- Check the appearance in a clear glass container (kerosene should be colorless).
- Smell the liquid (no unusual odor).
- Observe the flame while burning (no smoke, soot, or odor).
- Check the wick after the first filling (it should remain soft with no deposits).

ESTIMATING SAVINGS

A reason for investing in a kerosene heater is potentially saving money from a lower thermostat setting. Table 2-3 will help you in roughly estimating this savings. Your local utility or other fuel supplier may be able to help you with additional calculations. In Table 2-3 a potential 20 percent savings from lowering your thermostat from 70° F. (24 hours per day) to 60° F. (24 hours per day) is assumed. You will not achieve similar dollar savings if you go from 65° F. down to 55° F., but for this *rough* estimate you can use the 20 percent figure. Lower settings may allow water pipes to freeze. Note also that credit is not allowed for sleeping hours, nor for hours when the home is unoccupied. These savings may be captured by setting back the thermostat (manually or automatic) and have nothing to do with whether or not you purchase a kerosene heater.

Table 2-3 doesn't attempt to estimate the additional costs due to the central heating system operating less efficiently at lower thermostat settings (due to stop and go cycling). This may be roughly offset by the possibility of leaving the central system completely off (using the portable instead) for some weeks at the beginning and end of each heating season.

You will use the portable heater to maintain certain areas at 68° F. or higher. Children playing on the floor may be chilled while adults at a higher level in the room may be comfortable. The kids may prefer to be cooler, though, because they are more active. Elderly people may have trouble maintaining body temperature if the room temperature is below 70° F.

Table 2-4 will help you roughly estimate annual operating expenses for a portable kerosene heater. First, what size heater would you buy? Manufacturers usually recommend multiplying the area of the room by 28 BTU/hr. per square foot to obtain heater rating. Heater output is not very adjustable. Don't buy a heater that is too large. The excess heat will go to waste. Any dealer will quote a price for the size you need. Remember that list prices may be negotiable. Watch for sales or discounts. Get prices for wicks ($5-$20), batteries, and other maintenance items.

Table 2-3. Estimate of Potential Savings from Lower Thermostat Setting (courtesy of Cornell University Cooperative Extension).

Your Data	Sample Data		
	$1300	A.	Your last season's heating bill total (check with fuel supplier).
	$ 260	B.	Multiply: (A) by 0.20 (estimate 20 percent gross savings, see text).
	8½	C.	Number of sleeping hours per night.
	9½	D.	Number of hours per day when no one is home (away working, shopping, etc.).
	6.7	E.	Multiply: (D)×0.7 if you're in the house on weekends, if not then (D)×1.0.
	15.2	F.	Add: (C)+(E) (≅ hours per day when heater accounts for no savings).
	8.8	G.	Subtract: 24−(F) (possible hours per day for heater).
	0.37	H.	Divide: (G)÷24 (fraction available for heater operation).
	$ 96	I.	Multiply: (B)×(H) (potential gross savings due to heater).

Table 2-4. Estimated Annual Costs for Portable Kerosene Heater Operation (courtesy Cornell University Cooperative Extension).

Your Data	Sample Data		
	$180	A.	Heater cost: $ total (see text).
	$ 20	B.	Parts and service (wicks, batteries, etc.) See text: $/year.
	$ 1.30	C.	Fuel cost: $/gallon.
	0.058	D.	Fuel consumption: gallons/hr.
	8.7	E.	Operating hours per day (=Table 2-3, line G).
	150	F.	Days in heating season.
	75	G.	Estimated fuel use: multiply: $(D) \times (E) \times (F)$ = gallons/year.
	$ 98	H.	Estimated annual fuel cost: multiply $(C) \times (G)$ = $/year.
	$ 5	I.	Your time value (see text): $/hr.
	5	J.	Time required: minutes/gallon.
	$ 31.25	K.	Time cost: multiply: $(G) \times (I) \times (J) \div 60$ = $/year.
	$150	L.	Rough total annual cost: add: $(B) + (H) + (K)$ = $/year

Ask about service prices and policies. Don't forget to add taxes on all these purchases. Wicks should be replaced annually, but bad fuel will quickly necessitate replacement.

Check various suppliers for fuel costs. Specify that you intend to burn it in an unvented heater. Fuel consumption ranges for your size heater are usually given in the sales brochure, or divide heater rating (Btu/hr.) by 135,000 Btu/gallon to get fuel use in gallons per hour. The sample data were calculated using a heater turned down to 90 percent of full output (8700 Btu/hr. × 0.90 = 7830 Btu/hr., and 7830 ÷ 135,000 = 0.058 gallon/hr). For the number of days in the heating season, contact your fuel supplier or use your own experience.

What's your time worth to you? It's certainly not free, and it may be worth more to you than your regular hourly wage. Ask heater owners about the time required for cleaning and changing wicks. It also takes time to buy, transport, and store fuel and fill the heater with it. A very rough estimate might be 10 to 20 minutes per refueling (i.e., 5 to 10 minutes per gallon for a 2-gallon tank).

HEATER EFFICIENCY

Only one more piece of information is needed for cost comparisons, but it's a tough one to estimate. What is the *efficiency* of a portable kerosene heater?

Manufacturers commonly claim 99.9 percent efficiency. We generously assume this for the sample problem. Because any unvented heaters allows all combustion products to remain in the room, this claim seems obvious. The manufacturer's specifications for the heater rating is always larger than the heater output in Btu/hr. This is because burning 1 gallon of kerosene produces about 1 gallon of water, for which no heat value is allowed until it condenses. An engineer would calculate the efficiency of an unvented kerosene heater as output Btu/hr. divided by input Btu/hr. which equals about 94 percent. In some cases this added moisture may be welcome as it raises the indoor humidity in the vicinity of the heater. Other losses may reduce the amount of heat actually delivered to the room. Ventilation is sometimes recommended to include a window opening of 1 inch. Depending on outside conditions, the loss from this opening could actually exceed the maximum output of the heater. Although this is unlikely, an opening will always cause some loss.

Heater output is not very adjustable. When the lowest setting still produces more heat than is needed in the room, you can either lower the central heating still further, which may not be desirable, or waste some heat to another room or to the outdoors.

As the sample data shows, portable kerosene heaters may sometimes cost more than they save. The economic calculations are very sensitive to such factors as: the potential operating hours per day, heater size, fuel prices, and time requirements. Estimates of these inputs should be made carefully. The major problem is with estimated possible gross savings, as even small changes could make a large difference. Without fairly detailed engineering calculations it is impossible to better gauge potential savings from lowering thermostat settings even farther—say, from 70° F. down to 50° F. If we believed a manufacturer's claims of a potential 50 percent gross savings, then Table 2-3 would give a $240 per year credit for the heater.

HEATER PAYBACK

Payback is defined as the number of years of savings necessary to pay for an investment. The total heater cost is given in Table 2-4, line A. Net annual savings is the difference between gross potential savings and total annual cost (3I-4L). The expected payback is: total cost + net annual savings or calculate (4A) ÷ (3I-4L) = _____ years to pay for the cost of the heater. Obviously, if net annual savings are negative, you don't buy the heater. Although payback is a crude measure that ignores increasing fuel costs, it may be used to compare different energy investments. Your expected payback should be compared to similar numbers for weatherization and conservation investments. The lowest number is the fastest payback, which means it is the best energy investment.

You might also wish to compare portable electric costs. Initial cost is usually lower. Maintenance and time costs are negligible. The calculation for an equivalent amount of heat (fuel) is:

Annual Electricity cost =
($/kWH) × (gallons kerosene/
year) × (39.6 kWH/gallon) × efficiency.

For sample data using 6 cents per kilowatt-hour and 94 percent efficiency:

Annual electricity cost =
(0.06) × (75) × (39.6) × (.94) = $167.50.

SELECTIVE KNOWLEDGE

The best way to select the most appropriate portable kerosene heater is to understand the basic components and how they work together. This knowledge will help you choose a kerosene heater that will give you greatest efficiency with the least maintenance problems.

While not all kerosene heaters have identical parts, they do work on the same principles. A study of the typical components will give you valuable insight into selecting the most efficient heater for your needs.

WICK

A kerosene heater is an appliance in which kerosene is gasified by surface evaporation and burned. The amount of kerosene evaporated and heat generated can be increased in direct proportion to the area of the contact surface between the kerosene and air.

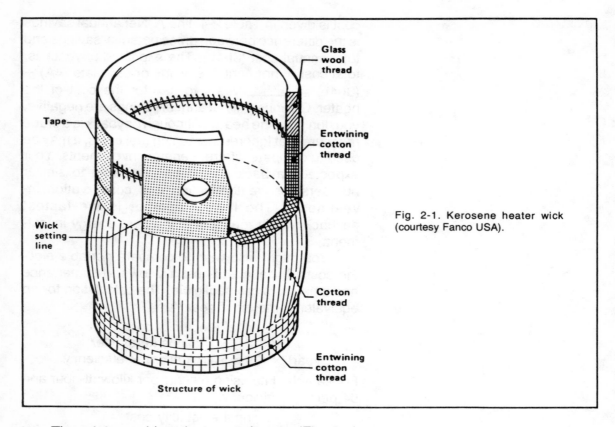

Fig. 2-1. Kerosene heater wick (courtesy Fanco USA).

Glass wool thread

Entwining cotton thread

Tape

Cotton thread

Wick setting line

Entwining cotton thread

Structure of wick

The *wick* used in a kerosene heater (Fig. 2-1) consists of many bundles of fine fibers and is designed to provide a large evaporation area. A wick consists of bundles of thin fibers and countless capillary tubes. The kerosene is drawn up from the tank into the combustion area by these capillary tubes.

If the kerosene becomes viscous or dirt and dust find their way inside the heater, the capillary tubes will become clogged. This will cause a deterioration in the drawing of the kerosene, and combustion will no longer be possible. A wick is formed by a fiberglass wick at the top and a cotton wick at the bottom sewn together.

Fiberglass Wick

The *fiberglass wick* is made of special glass fibers (60 percent) and carbon fibers (30 percent) with consideration given to its strength and workability. This wick causes the kerosene to vaporize from the surface of the very thin glass fibers and sends the kerosene vapor to the combustion section. Combustion results when ignition takes the form of a direct flame (Fig. 2-2). The flame is fed by the kerosene vapor without direct contact with the wick.

Cotton Wick

The *cotton wick* is made of cotton fibers with excellent moisture absorption properties. It lifts the kerosene in the tank to the fiberglass wick using a capillary tube effect. In order that the wick height may be adjusted smoothly, a special weaving method is used for the bottom of the wick where threads running crosswise are not employed. This increases the wick's elasticity.

For initial ignition, more than 20 minutes should be allowed to elapse after the kerosene has been supplied before actual ignition. The wick lifts the kerosene to the combustion section using the capillary tube effect. Depending on the size of the capillary tubes, there may be some difference in the time taken for the kerosene to reach the top of the wick. If ignition is attempted before the kerosene has soaked sufficiently into the wick, many air holes are formed in the middle of the wick. This downgrades the kerosene suction effect (Fig. 2-3). When kerosene is not drawn

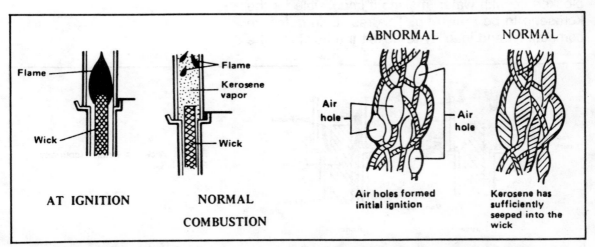

Fig. 2-2. Wick flame (courtesy Fanco USA).

Fig. 2-3. Capillary action (courtesy Fanco USA).

31

up properly, the flame comes into contact with the wick. The temperature of the wick rises to an abnormally high level, and the vaporization ability of the kerosene drops. Kerosene, which has once been vaporized, now comes into contact with the wick frame or air tube. It cools off and results in the formation of tar on the wick. Consequently, the kerosene's drawing ability gradually deteriorates so that combustion is no longer possible. Wick height adjustment can no longer be made.

If kerosene of a different quality or of a quality liable to undergo change or oils with a higher viscosity than kerosene are mixed in with the kerosene, the drawing ability deteriorates. The capillary tubes block up and, as in the previous situation, combustion cannot be produced. The wick height adjustment can no longer be made.

If kerosene is mixed with gasoline or benzene or an oil with a low level of viscosity is used, the petroleum product will be drawn up before the kerosene to create the danger of uneven flames or explosion. Also, air holes will be formed in the middle of the wick, resulting in lessened drawing ability, defective combustion, and the inability to adjust the height of the wick.

If kerosene containing water is used and the water is drawn up, the capillary tubes will become blocked up with water, making it impossible for the kerosene to be drawn up. This results in defective combustion and inability to adjust the height of the

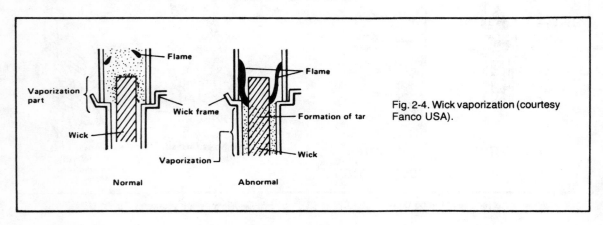

Fig. 2-4. Wick vaporization (courtesy Fanco USA).

wick. Also, if the wick height is reduced for use, the flame comes in contact with the wick. This results in the formation of tar on the wick, no combustion, and the inability to adjust the height of the wick.

Figure 2-4 is a cross section of the wick's vaporization position.

BURNER UNIT

The *burner unit* serves to mix the kerosene vapor after it has been vaporized from the wick, with the oxygen in the air for combustion (Figs. 2-5 and 2-6). In a kerosene heater the burner unit supplies air for combustion using the draft effect of the hot air current.

Portable kerosene heater burner units can be divided into two types by their structure: *Radiant* type (Fig. 2-7) with a glass outer cylinder and *convective* type (Fig. 2-8) with a metal outer cylinder.

Wear gloves when disassembling a glass burner unit assembly. If you touch the inner tube with bare hands, the part touched may be blackened when burning. The burning efficiency will deteriorate.

The burner unit should be set properly and securely on the wick frame and air tube. If the burner unit is not set properly, the air supplied by the draft will become unstable. An irregular eddy motion will be caused in the kerosene vapor and combustion gases, and the flame formations will be outside the combustion area. This will result in carbon monoxide or odors and soot and black smoke.

Any soot formed on the inner and middle tubes of the burner tube must be promptly cleaned. The small

Fig. 2-5. Portable kerosene heater burner units (courtesy Kero-Sun, Inc.).

Fig. 2-6. Once the cover is removed, the burner unit is displayed (courtesy Kero-Sun, Inc.).

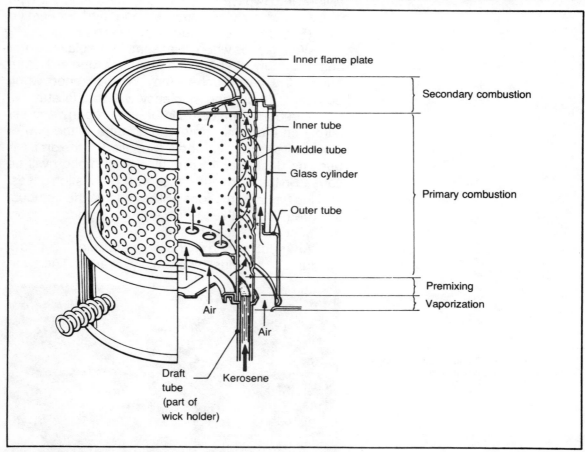

Inner flame plate

Secondary combustion

Inner tube

Middle tube

Glass cylinder

Primary combustion

Outer tube

Premixing

Vaporization

Air

Air

Draft
tube
(part of
wick holder)

Kerosene

Fig. 2-7. Radiant burner (courtesy Fanco USA).

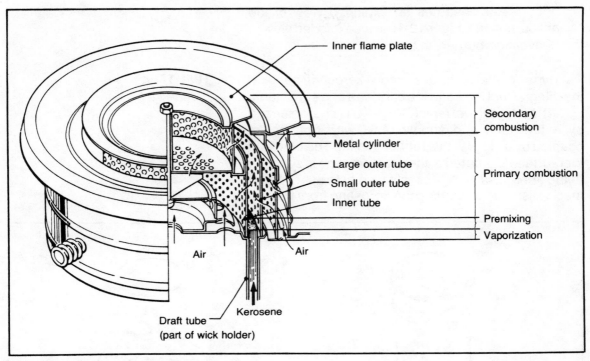

Inner flame plate

Secondary combustion

Metal cylinder

Large outer tube

Primary combustion

Small outer tube

Inner tube

Premixing

Vaporization

Air

Air

Kerosene

Draft tube
(part of wick holder)

Fig. 2-8. Convective burner (courtesy Fanco USA).

holes in the inner and middle tubes have a pitch that allows the air and the kerosene vapor to be mixed in the most effective way. If soot forms, on just one of these holes, the whole balance is upset and imperfect combustion results.

A burner unit that has been deformed or mis-shaped must not be used. If the inner tube, middle tube, outer tube (including glass), flame-extending plate, or radiator coil is deformed, turbulence arises in the air supply, causing imperfect combusion. Radiant heaters must not be operated with cracked or broken glass.

No attempt must be made to redesign the burner unit. Small holes have been made in the inner tube and middle tube so that the overall combustion balance is attained most effectively. Any attempt to change the hole diameter or remove parts will result in the loss of the overall combustion balance and imperfect combustion. Excessive carbon monoxide may be created.

Figure 2-9 illustrates an exploded view of the radiant burner unit. Figure 2-10 is an exploded view of the convection burner unit.

Two types of fuel tanks are used in kerosene heaters: the integral fuel tank and the cartridge and sump tank.

FUEL TANK

The *integral fuel tank* (Fig. 2-11) is characterized by a fuel tank and supply tank, with both tanks coupled by a slit or with a rubber gasket. The constant fuel surface is created at the slit. Kerosene is supplied to the fiberglass wick. This slit also acts as a control mechanism for preventing the kerosene from flowing

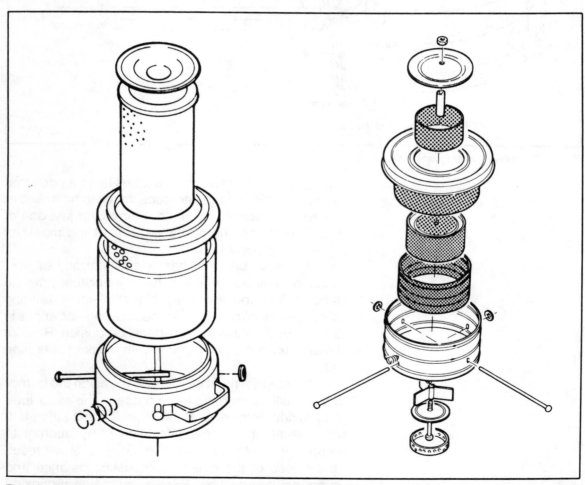

Fig. 2-9. Exploded view of radiant burner unit (courtesy Fanco USA).

Fig. 2-10. Exploded view of convection burner unit (courtesy Fanco USA).

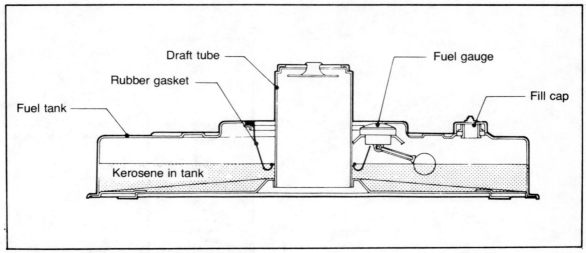

Fig. 2-11. Integral fuel tank (courtesy Fanco USA).

out when the heater has been tipped over. Any leakage in the gaskets of the fuel cap assembly or fuel gauge will make it impossible to control the constant kerosene surface and will cause the kerosene to escape when the heater is tipped over.

In the *cartridge and sump tank* system (Fig. 2-12), the constant kerosene surface for the kerosene supplied from the cartridge tank is created by the burner tank coupling and fuel cap assembly of the cartridge tank. Any leakage of air into the fuel tank coupling will make it impossible to create the constant kerosene surface and cause all the kerosene to flow out from the cartridge tank.

The settings of the cartridge tank and burner tank are shown in Fig. 2-12. The pin pushes up the valve holder and O-ring. The airtight sealing of the cartridge tank is released, and the kerosene flows out into the burner tank. This flow is automatically cut off as the bottommost level of the fuel supply orifice lid has been reached. This valve unit works according to *Torricelli's principle*. When the cartridge tank is removed, the spring is activated. The O-ring is sandwiched between the valve holder and the fuel cap assembly. The flow of kerosene is then blocked off, and the cartridge tank can be carried away.

Figure 2-13 shows the fuel supply coupling unit.

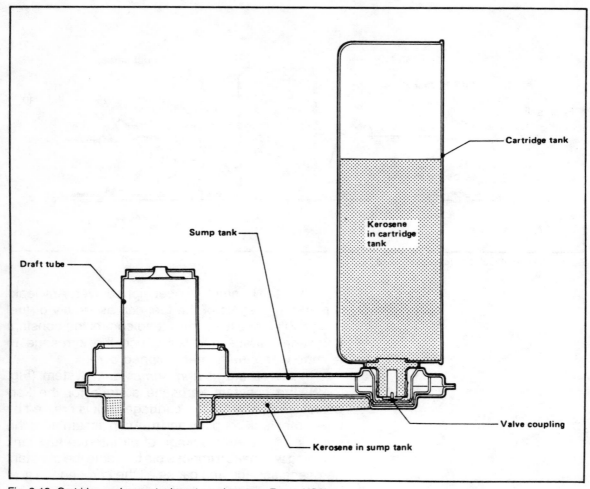

Fig. 2-12. Cartridge and sump tank systems (courtesy Fanco USA).

The *ignition mechanisms* in radiant and convection heaters are slightly different. Figure 2-14 illustrates the ignition mechanism for a radiant burner unit.

When the ignition knob is pushed down, the lever rises at the same time the burner unit is raised. The igniter coil rotates. When it approaches the wick, ignition results. When the end of the igniter coil does not touch the wick, no ignition results. If the wick height is not correct, adjust it for immediate ignition.

If you have problems with the automatic lighting mechanism on the radiant heater, first replace the batteries. If the glow of the igniter is not restored, the

IGNITION MECHANISM

igniter coil may be burned out and require replacement. This should be done on a completely cool heater. First allow auto shutoff to snap shut by jarring the heater body. Remove the batteries to prevent possible burns. Release the front grille and remove the chimney. Then push the ignition lever to expose the igniter coil. Hold the socket with your free hand (Fig. 2-15). Remove the igniter by pushing in and twisting in a counterclockwise direction to release it from its socket. Insert the replacement igniter, turn clockwise, and lock in position. Reinstall batteries and chimney, then close the grille. Slide the safety shutoff lever fully to the left to rest the auto shutoff mechanism.

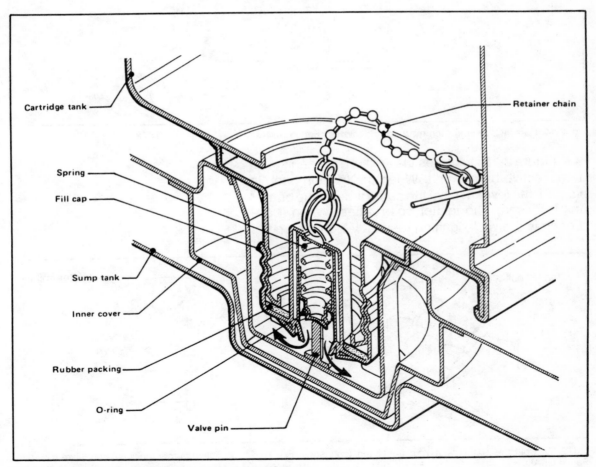

Fig. 2-13. Fuel supply coupling unit (courtesy Fanco USA).

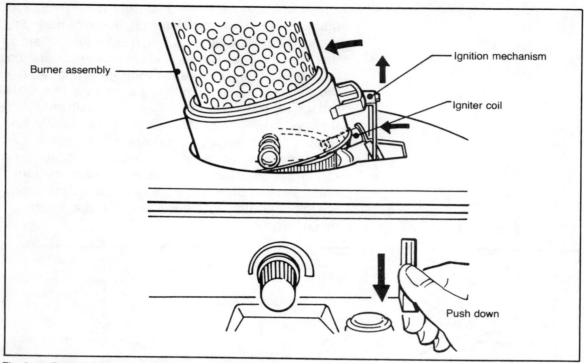

Fig. 2-14. Radiant burner ignition mechanism (courtesy Fanco USA).

Figure 2-16 illustrates the ignition mechanism on a convective burner unit. When the ignition knob is pushed, the lever rises at the same time as the burner unit is raised. The igniter coil rotates. When it approaches the wick, ignition results. When the end of

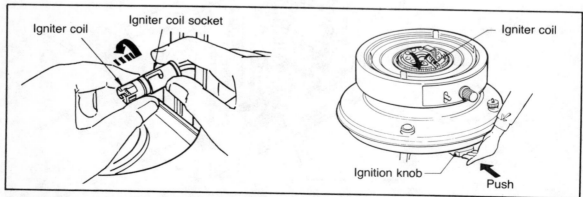

Fig. 2-15. Hold the socket with your free hand (courtesy Fanco USA).

Fig. 2-16. Convective burner ignition mechanism (courtesy Fanco USA).

the igniter coil does not touch the wick, no ignition results. If the wick height is not correct, adjust it for immediate ignition.

If you have problems with the automatic lighting mechanism on the convection burner unit, first replace the batteries. If the proper glow of the igniter is not restored, the igniter coil may be burned out and need replacing. As with the radiant burner coil, this should be done on a completely cool heater. Allow the auto shutoff to snap shut by jarring the heater body. Remove the batteries to prevent possible burns. Remove the top plate by spreading a wire handle that holds it in place. Remove the outer cover by removing four screws. Lift the burner unit out of the heater. Push the ignition knob to expose the igniter coil. Hold the socket in your free hand and remove the igniter by pushing in and twisting counterclockwise to release it from its socket (Fig. 2-17). Insert the replacement igniter. Turn clockwise to lock it into position. Reinstall the batteries, outer cover, burner, top plate, and carrying handle.

Figure 2-18 illustrates the igniter coil used on most radiant and convection portable kerosene heaters. The temperature of the filament rises to 1470° F. for 0.3 seconds. Rush current is 3.3 volts with 1.6 amps. CAUTION: it is possible that the platinum-rhodium alloy filament will burn out when it is always exposed to the combustion gases after ignition. The

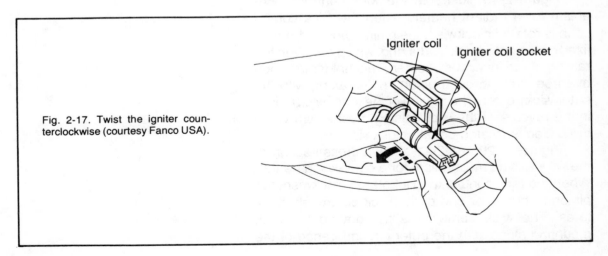

Igniter coil

Igniter coil socket

Fig. 2-17. Twist the igniter counterclockwise (courtesy Fanco USA).

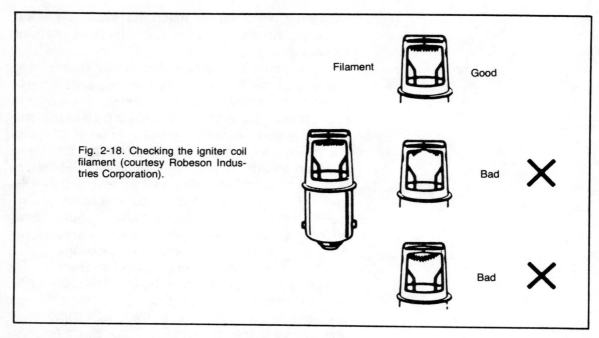

Fig. 2-18. Checking the igniter coil filament (courtesy Robeson Industries Corporation).

filament itself is a thin wire and, if deformed, may break.

Figure 2-18 shows normal and abnormal filament coils.

WICK CONTROL MECHANISM

The wick within a kerosene heater must be raised and lowered as needed. This is done by the *wick control mechanism*.

Figure 2-19 illustrates the wick control mechanism for a radiant heater. When the wick control knob is rotated clockwise, the pinion moves the rack fixed on the outer wick holder ring, which results in the raising of the inner wick holder. This holder is joined by three rivets through oblong holes along with the outer wick holder. The motion of the inner wick holder is the reverse of the wick raising mechanism when the pinion is rotated counterclockwise.

Figures 2-20 and 2-21 show a cross section of the wick raising mechanism for a convection heater. When the wick control knob is rotated clockwise, the pinion at the other end of the wick control shaft rotates. The wick frame rises by rotating the rack mounting along with the outer circumference of the wick frame. This results in the wick, which is mounted through the wick frame, being lifted up at the same time. When the wick control shaft is rotated counterclockwise, the energy stored in the torsion spring is retained by the latch that is coupled with the ratchet and shutoff device. The torsion spring is attached to the shaft at one end and to the wick control shaft at the other (Fig. 2-22).

When the wick control knob is rotated in the counterclockwise direction (Fig. 2-23) the relationship between the arm and gear is maintained by the bolt secured to the wick control knob and the unit attached to the ratchet. The pinion rotates in the counterclockwise direction with the energy stored in the torsion spring.

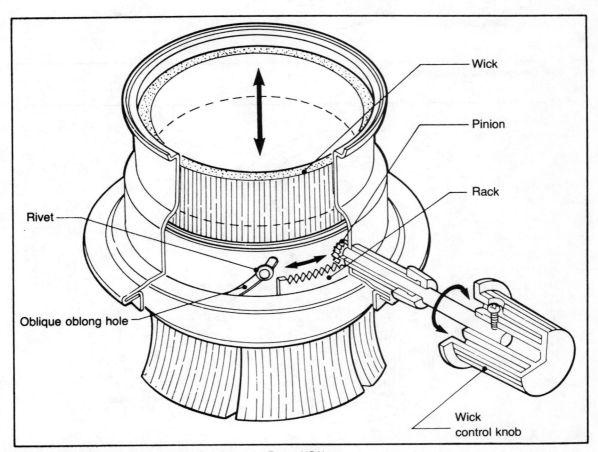

Fig. 2-19. Radiant wick control mechanism (courtesy Fanco USA).

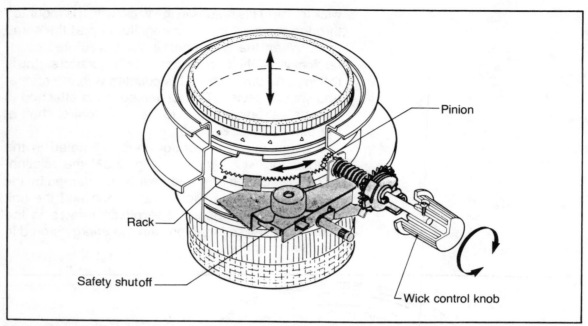

Fig. 2-20. Convective wick control mechanism (courtesy Fanco USA).

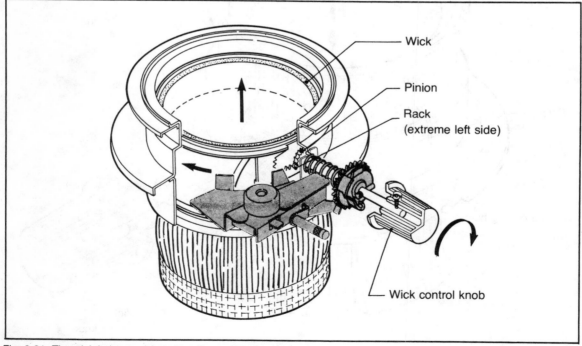

Fig. 2-21. The wick is lowered (courtesy Fanco USA).

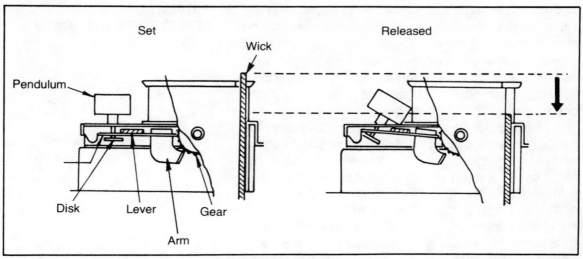

Fig. 2-22. Control mechanism set and released (courtesy Fanco USA).

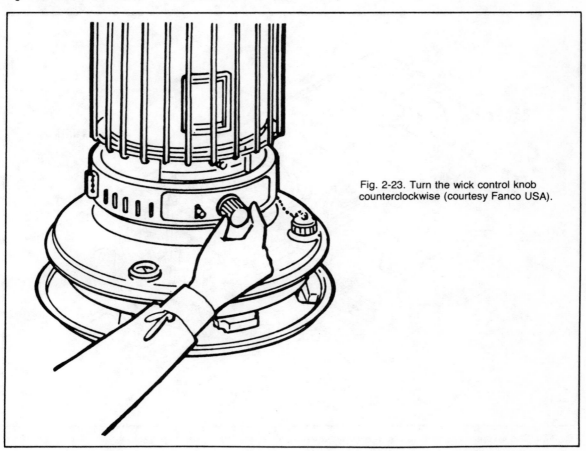

Fig. 2-23. Turn the wick control knob counterclockwise (courtesy Fanco USA).

There are two types of automatic extinguishing devices used in the majority of kerosene heaters: the shutter type and the drop type.

With the *shutter automatic extinguishing mechanism* (Fig. 2-24), the pendulum starts swinging if bumped or jolted. The slide moves quickly and covers the wick. This extinguishes the heater.

In the drop type (Fig. 2-25), a shock starts the pendulum swinging. The wick control shaft ratchet and wick control knob rotate counterclockwise by the force of the torsion spring after the latch has been released from the ratchet. This results in a quick drop of the wick and puts out the flame.

Never attempt to override the automatic extinguishing mechanism. It is designed to increase kerosene heater safety and protect your heater and

AUTOMATIC EXTINGUISHING MECHANISM

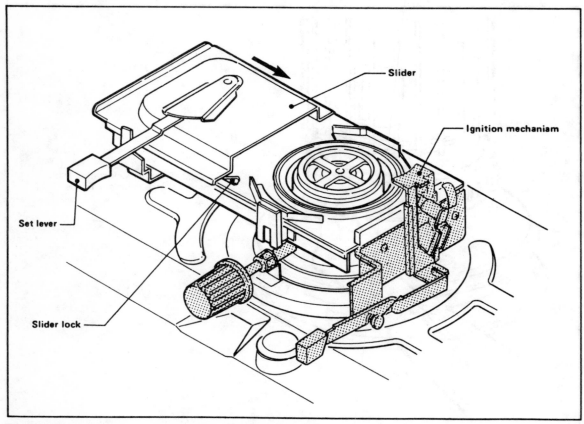

Fig. 2-24. Shutter automatic extinguisher (courtesy Fanco USA).

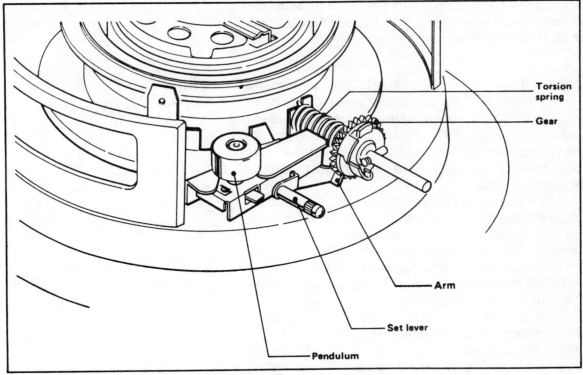

Torsion spring

Gear

Arm

Set lever

Pendulum

Fig. 2-25. Drop automatic extinguisher (courtesy Fanco USA).

home from spilling fuel that would be lead to a fire (Figs. 2-26 and 2-27).

Those are the major components of the radiant and convection portable kerosene heaters. By understanding the separate components, their opera-

Fig. 2-26. Checking the automatic extinguisher mechanism (courtesy Kero-Sun, Inc.).

Fig. 2-27. Make sure there are no carbon deposits on the extinguisher ring (courtesy Kero-Sun, Inc.).

tions, and variations, you can select the most appropriate and efficient unit for your heating needs (Figs. 2-28 through 2-51).

Another important aspect of selecting a kerosene heater is choosing the size that will best fit your heating needs. As you have learned, kerosene heaters are not as adjustable as other heaters. They cannot be turned down to "low" without the adverse effects of increasing pollutants. Take care to select the most appropriate size.

What size heater should you buy? The most important consideration is the size of the room to be heated. Obviously, a 12-by-12-foot room doesn't need as large a portable kerosene heater as a 19-by-24-foot room. Another factor in room sizing is the height of the ceiling. An 8-foot ceiling requires less heat than a 12-foot cathedral ceiling room.

Another important factor in sizing your heater is the tightness of the house construction. Does the home allow a great deal of outside air to infiltrate the home through cracks around doors and windows? Is the insulation adequate? A tight home will minimize heat loss through walls and openings such as doors and windows. This heat loss must be estimated when choosing your heater.

The outdoor temperature is another important factor. If the typical winter day averages 20° F. outside, your heating requirements are going to be about a third less than if the typical temperature is around 0° F.

The fourth factor is your desired indoor temperature. Heating a home to 72° F. is more costly than heating it to 69° F. It also requires a larger heating system.

With an outside temperature of 0° F., a 9600 Btu portable kerosene heater will serve a 342-square-foot room (18 by 18 foot), 10500 Btus can warm a 360-square-foot room (18 by 20 foot), and 20000 Btus are need to heat a 676-square-foot room (26 by 26 foot).

With an outside temperature of 20° F., a 9600

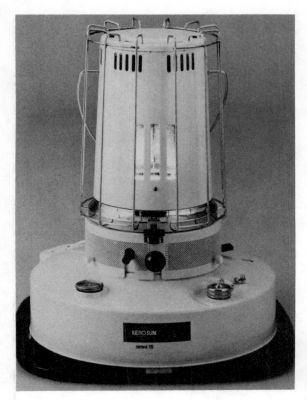

Fig. 2-28. Convection kerosene heater
(courtesy Kero-Sun, Inc.).

Fig. 2-29. Convection kerosene heater
(courtesy Kero-Sun, Inc.).

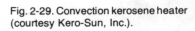

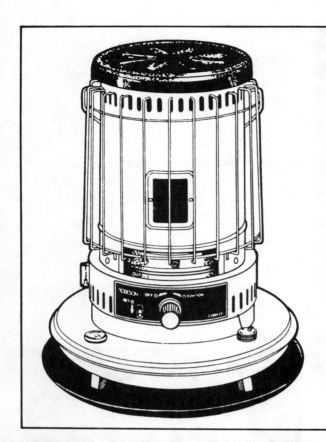

Fig. 2-30. Convection kerosene heater (courtesy Robeson Industries Corporation).

Fig. 2-31. Radiant kerosene heater (courtesy Kero-Sun, Inc.).

Fig. 2-32. Radiant kerosene heater (courtesy Kero-Sun, Inc.).

Fig. 2-33. Radiant kerosene heater (courtesy Kero-Sun, Inc.).

Fig. 2-34. Radiant kerosene heater (courtesy Teknika Electronics Corporation).

Fig. 2-35. Radiant kerosene heater (courtesy Teknika Electronics Corporation).

Fig. 2-36. Radiant kerosene heater (courtesy Teknika Electronics Corporation).

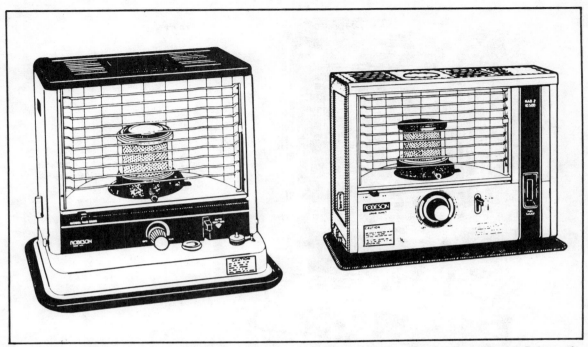

Fig. 2-37. Radiant kerosene heater (courtesy Robeson Industries Corporation).

Fig. 2-38. Radiant kerosene heater (courtesy Robeson Industries Corporation).

Fig. 2-39. Permanent kerosene heater (courtesy Kero-Sun, Inc.).

Fig. 2-40. Underwriters Laboratories listed heater (courtesy GLO International Corporation/Toyokuni).

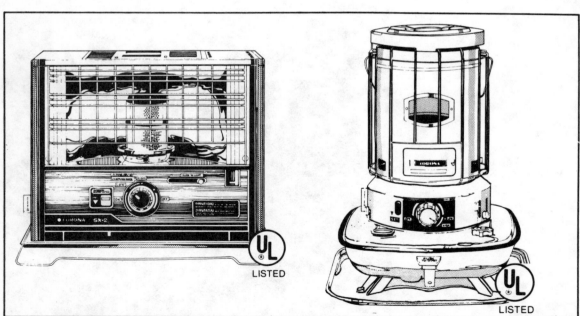

Fig. 2-41. Underwriters Laboratories listed heater (courtesy GLO International Corporation/Corona).

Fig. 2-42. Underwriters Laboratories listed heater (courtesy GLO International Corporation/Corona).

Fig. 2-43. Underwriters Laboratories listed heater (courtesy GLO International Corporation/Toyokuni).

Fig. 2-44. Underwriters Laboratories listed heater (courtesy GLO International Corporation/Toyokuni).

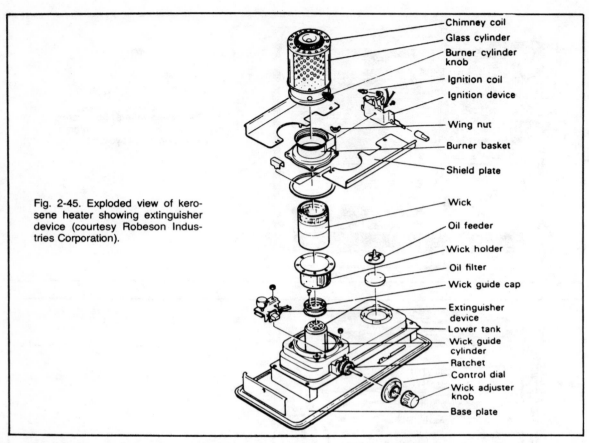

Fig. 2-45. Exploded view of kerosene heater showing extinguisher device (courtesy Robeson Industries Corporation).

Chimney coil
Glass cylinder
Burner cylinder knob
Ignition coil
Ignition device
Wing nut
Burner basket
Shield plate
Wick
Oil feeder
Wick holder
Oil filter
Wick guide cap
Extinguisher device
Lower tank
Wick guide cylinder
Ratchet
Control dial
Wick adjuster knob
Base plate

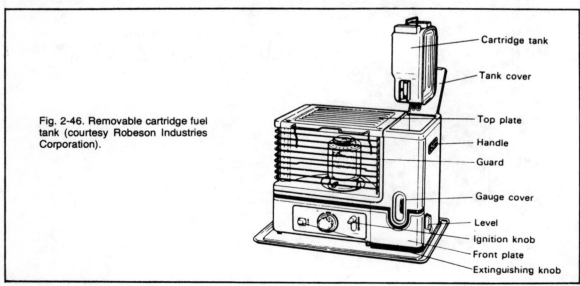

Fig. 2-46. Removable cartridge fuel tank (courtesy Robeson Industries Corporation).

Cartridge tank
Tank cover
Top plate
Handle
Guard
Gauge cover
Level
Ignition knob
Front plate
Extinguishing knob

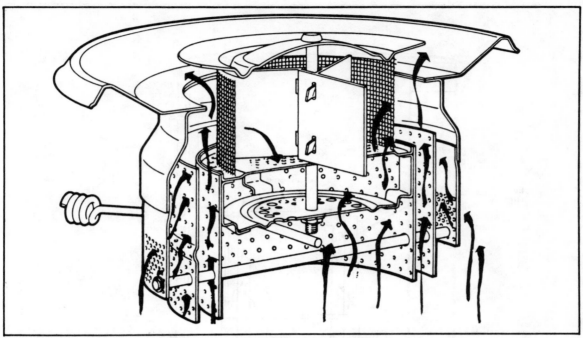

Fig. 2-47. Cross section of kerosene heater burner (courtesy Kero-Sun, Inc.).

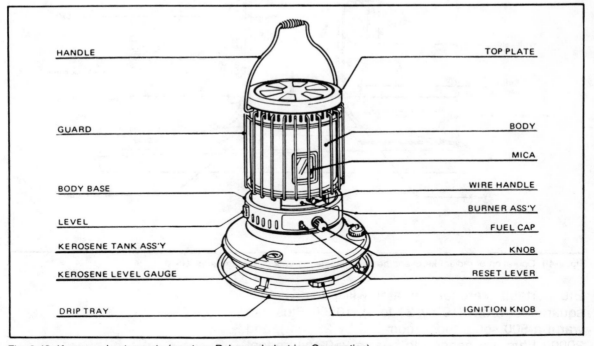

HANDLE

TOP PLATE

GUARD

BODY

MICA

BODY BASE

WIRE HANDLE

LEVEL

BURNER ASS'Y

FUEL CAP

KEROSENE TANK ASS'Y

KNOB

KEROSENE LEVEL GAUGE

RESET LEVER

DRIP TRAY

IGNITION KNOB

Fig. 2-48. Kerosene heater parts (courtesy Robeson Industries Corporation).

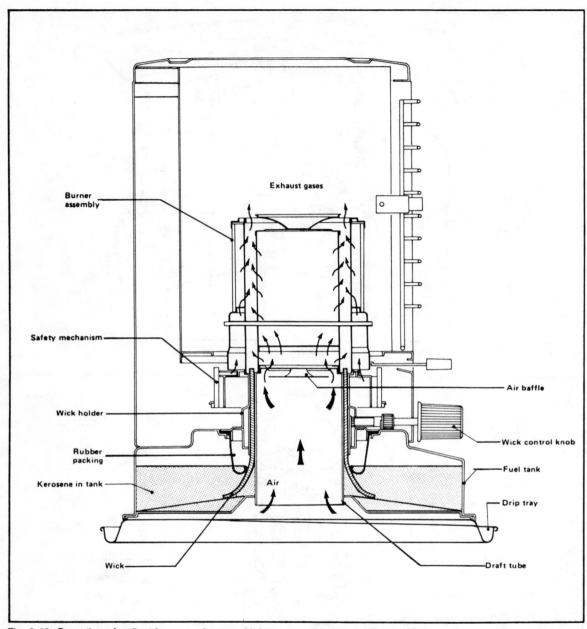

Fig. 2-49. Operation of radiant kerosene heater with integral tank (courtesy Fanco USA).

Btu portable kerosene heater will serve a 450-square-foot room (19 by 24 foot), 10500 Btus can warm a 500-square-foot room (20 by 25 foot), and 20000 Btus are needed to heat a 960-square-foot room or area (30 by 32 foot). An 8-foot ceiling is assumed in each of these examples.

By comparison, a 1500-watt portable electric heater generates about 5000 Btu/hr.

As you estimate the size of your required portable kerosene heater, the amount of insulation in your home is an important factor in the amount of heat

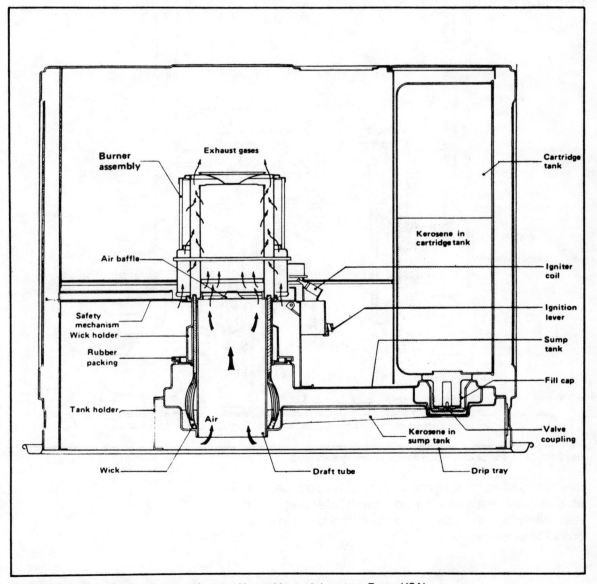

Fig. 2-50. Operation of radiant kerosene heater with cartridge tank (courtesy Fanco USA).

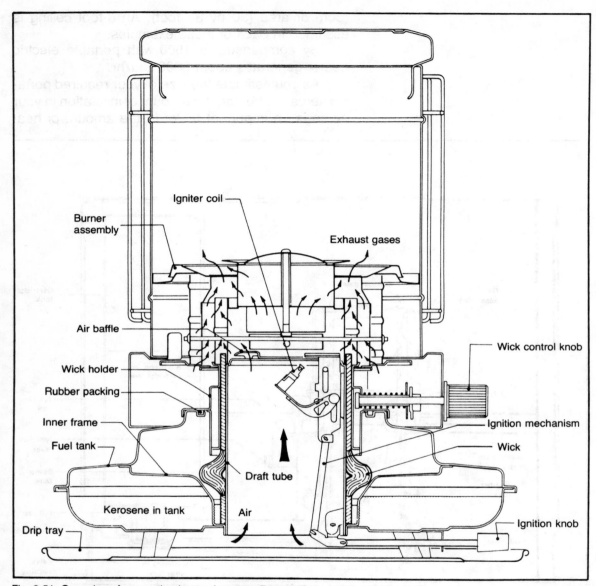

Fig. 2-51. Operation of convection heater (courtesy Fanco USA).

needed to keep your home comfortable. You may be able to purchase a smaller portable heater and use it less often by first making sure that your home is properly insulated.

Chapter 3

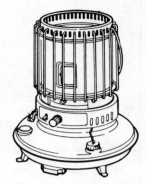

Using Your Kerosene Heater

YOU'VE SELECTED THE MOST APPROPRIATE KER-osene heater for your needs, and you're ready to put it to work. The first step is to gather all the paper-work that goes with the kerosene heater: operations manual, owner's manual, warranty or guarantee, and sales slip. Keep them all together, preferably in an envelope where they can be referred to as you need them.

OWNER'S MANUAL

The day of the one-page mimeographed product owner's manual is just about gone. Today, most pro-ducts requiring instructions include well-illustrated and somewhat clearly written manuals on setup, op-eration, maintenance, and cautions. Because ker-osene heaters are more complicated than other heaters, the owner's manuals are often better and more useful.

Information in this chapter is not intended to supersede the owner's manual that comes with your heater, but rather to supplement it with facts and illustrations that may help you better understand and enjoy your purchase.

UNPACKING YOUR HEATER

Figures 3-1 and 3-2 illustrate how radiant and con-vective kerosene heaters are typically packed for shipment to the consumer.

Open the box, take the heater out of the card-

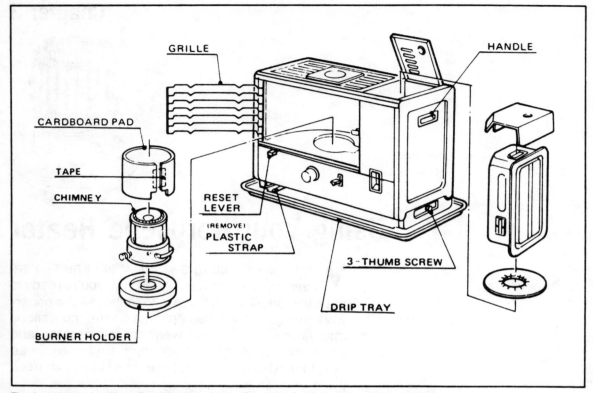

GRILLE

HANDLE

CARDBOARD PAD

TAPE

CHIMNEY

RESET LEVER

(REMOVE)
PLASTIC STRAP

3 - THUMB SCREW

DRIP TRAY

BURNER HOLDER

Fig. 3-1. Unpacking the radiant kerosene heater (courtesy Robeson Industries Corporation).

board case, and remove pads. Check in the packing for batteries, manuals, literature, and guarantees.

The heater itself may have additional packing material within or attached to it. Lift the guard and remove pads or packing.

If included, remove the pendulum hook of the extinguisher device (Fig. 3-3). Unless the pendulum hook is removed on some models, the extinguisher cannot be set.

Save the cardboard case and all inside packing as a storage box.

Lay out the pieces of the kerosene heater on a hard, level surface such as a counter or tabletop. Don't try to assemble your heater on a rug where small parts and screws can become lost. Check the inventory list and exploded view of your heater in the owner's manual to make sure that all parts are included. If anything seems to be missing, go back

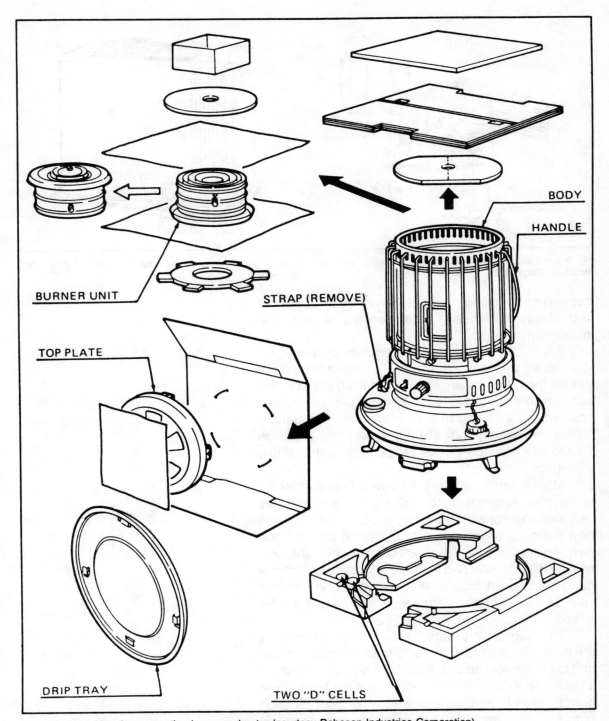

BODY

HANDLE

BURNER UNIT

STRAP (REMOVE)

TOP PLATE

DRIP TRAY

TWO "D" CELLS

Fig. 3-2. Unpacking the convection kerosene heater (courtesy Robeson Industries Corporation).

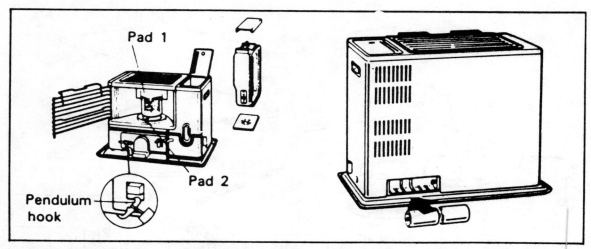

Pad 1

Pad 2

Pendulum
hook

Fig. 3-3. Release the pendulum hook (courtesy Robeson Industries Corporation).

Fig. 3-4. Install batteries (courtesy Robeson Industries Corporation).

through the packing material. Never try to assemble and use your kerosene heater unless all parts are installed properly.

If your kerosene heater requires batteries for ignition, install them now (Fig. 3-4). Depending on the model, the batteries are installed in the base or at the back of the unit. Polarity is important. Batteries have a positive (+) and negative (−) terminal. The battery holder will usually indicate which way the batteries should be installed—positive to positive and negative to negative.

Most heaters use dry batteries that cannot be recharged. Attempting to recharge these batteries can be dangerous as they may explode. Also, install new batteries together rather than one old and one new. The old one may drain off the current of the new one, and you'll have *two* inefficient batteries. Remove dead batteries at once as they can leak and damage your heater. Remove batteries when the unit will be stored or unused for any length of time.

The burner unit rests on the ring that encircles the wick (Fig. 3-5). Once you've made certain that the unit is on a level surface, install the burner and rotate it back and forth until it is in position. If the burner unit is not properly aligned, the heater will not function correctly and smoking may occur. In some models

Fig. 3-5. Burner unit on the wick ring (courtesy Robeson Industries Corporation).

you can confirm the lowering down of the wick by pushing down on the extinguishing knob.

When you're satisfied that all packing material is removed, that all parts are installed properly, and that any batteries required are correctly in place, it's time to fuel your kerosene heater.

FUELING YOUR HEATER

One of the most important and critical steps in preparing your kerosene heater for use is fueling it.

Because kerosene is not a common household substance, we first need to look at what it is and how it works.

Kerosene is a combustible, nonflammable liquid obtained by distilling petroleum. It is related to gasoline and other fuel oils, *but should never be used interchangeably with other petroleum products*. This caution will be repeated many times in this book and in most owner's manuals for kerosene heaters. People still think that kerosene is just another type of gasoline, which it isn't, and try to fuel their kerosene heater with gas. The results can be explosive.

Not all kerosene is alike. The type to be used in kerosene heaters is called "grade K-1 low sulfur." It is top grade kerosene with a low amount of sulfur—0.04 percent or less, by weight. K-2 or 2-K kerosene has more sulfur in it. What's the difference? The burning of sulfur produces sulfur dioxide, which is a toxic, irritant gas that can cause breathing difficulty espe-

65

cially in a closed room and most especially to those with asthma or other respiratory problems.

How can you tell the difference between K-1 and K-2 kerosene? K-1 kerosene is usually clear or colorless while K-2 is yellow because of the higher sulfur content. The specific gravity of kerosene is about 0.8, and its ignition point is more than 100° F. To distinguish kerosene from other fuels, take ½ teaspoon of fuel and bring it close to a lit match. If the fuel has a low ignition point, such as gasoline or thinner, it will burn. If it is pure K-1 low sulfur kerosene, it will not (Fig. 3-6).

Low grade kerosene can also reduce the life of the wick and cause caking of excess sulfur and smoking. The greatest hazard here is that the caked wick will hamper the action of the extinguisher that puts out the wick's flame if the kerosene heater is accidentally tipped over.

Kerosene can also deteriorate if it is stored for more than one heating season, stored in a sunlit place or an area of high temperature, or mixed with other fuels including bad kerosene. Deteriorated kerosene has a light brownish or dark yellow color. Kerosene will easily deteriorate in any of these cases, especially if it is stored in a plastic container. Kerosene in a plastic container can become unsuitable fuel in just two weeks.

Make sure that the kerosene you use to fuel your kerosene heater has not been mixed with anything

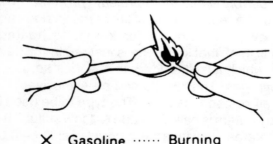

X   Gasoline ······ Burning

O   Kerosene ······ Not burning

   (K-1 LOW SULPHUR)

Fig. 3-6. Kerosene liquid does not burn (courtesy Robeson Industries Corporation).

else—gasoline, diesel, fuel oil, etc. Don't use the same can used for other fuels unless it has been thoroughly rinsed out with kerosene and relabeled.

REFUELING PROCEDURES

Some kerosene heaters have removable cartridge tanks while others have built-in fuel tanks. In either case, do not attempt to refuel your heater until it is turned off and has cooled. Refuel outside rather than inside your home as spilled kerosene can be more dangerous than contained kerosene, especially kerosene spilled on a rug or carpet as it acts as a wick.

Figure 3-7 illustrates how to remove a cartridge tank from a kerosene heater and check the fuel level (Fig. 3-8). Figure 3-9 shows how to check the fuel level in units with built-in tanks.

Many kerosene heaters have siphons included. If not, purchase one from the dealer when you buy your heater. It is difficult and unsafe to transfer fuel from your kerosene container to the heater tank without a siphon.

To use the siphon, tighten the airtight cap on the top of the siphon and insert the straight hose into the

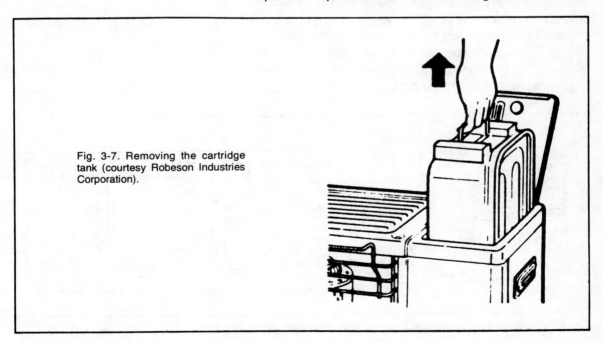

Fig. 3-7. Removing the cartridge tank (courtesy Robeson Industries Corporation).

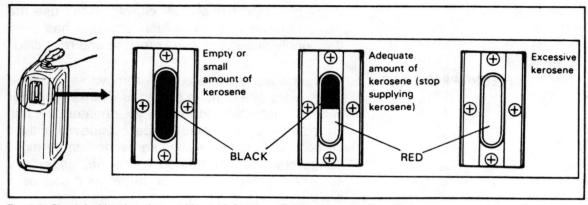

Fig. 3-8. Check fuel level on the cartridge tank (courtesy Robeson Industries Corporation).

kerosene container. Insert the flexible hose through the opening of the cartridge tank or built-in tank (Figs. 3-10 and 3-11). Squeeze the bulb repeatedly and refuel by watching the fuel gauge. In some models, fill until the indicator reaches the "F" or "full" mark. In others, only fill until it reaches a colored area on the indicator. *Do not overfill*.

Fasten the tank cap securely. If your heater is a cartridge tank type, install the tank in the heater. Make sure that it sits properly within the unit. Wipe up any spilled kerosene thoroughly and store the rag in a closed container such as an empty coffee can with plastic lid. Again, kerosene burns through capillary action. The fumes burn. Danger is minimized when rags are kept in a closed can.

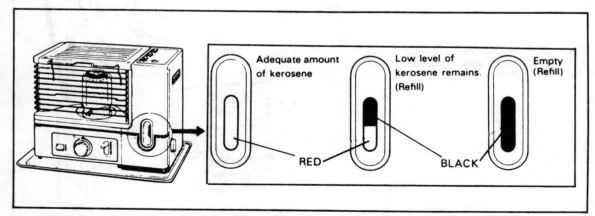

Fig. 3-9. Check fuel level on the built-in tank (courtesy Robeson Industries Corporation).

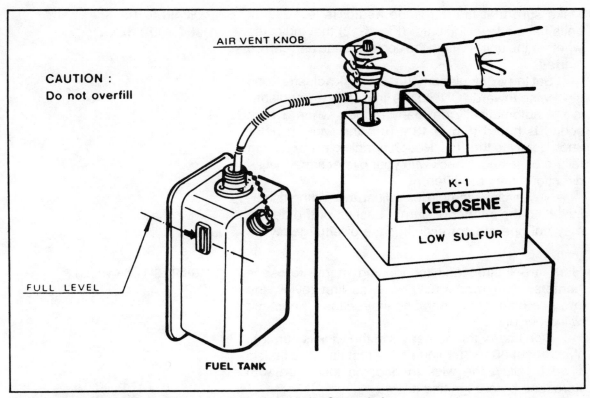

Fig. 3-10. Fueling the cartridge tank (courtesy Robeson Industries Corporation).

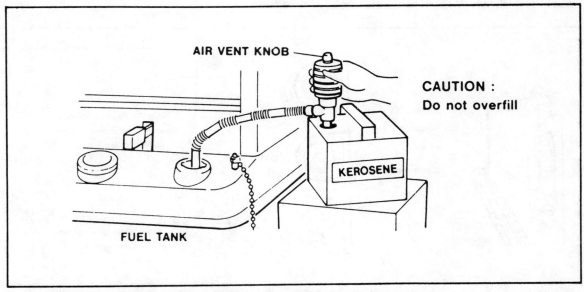

Fig. 3-11. Fueling the built-in tank (courtesy Robeson Industries Corporation).

Make sure that the kerosene heater is level. Most units have a level indicator (Fig. 3-12) that indicates whether the unit is within the correct range and can be started.

Set the extinguisher. Turn the wick adjuster knob clockwise toward "ON" to the limit. The extinguisher is set automatically on many models when a click sound is heard (Fig. 3-13). Turn the wick adjuster knob gently to the limit. Rough handling may result in failure of setting. Check with your owner's manual as directions may be different.

Never disassemble or adjust the extinguisher, including the interior or burner basket, or the device may not operate correctly in case of emergency.

PREPARING TO LIGHT YOUR HEATER

Here are instructions for igniting many kerosene heaters. Your model may vary, so first review the owner's manual or operating instructions that came with your unit.

Don't carry the heater while the wick is burning. Wait about 20 to 30 minutes the first time the tank is filled to allow the wick to become saturated with kerosene before igniting it.

First, engage the safety shutoff reset lever by moving it downward. This balances a pendulum

LIGHTING YOUR HEATER

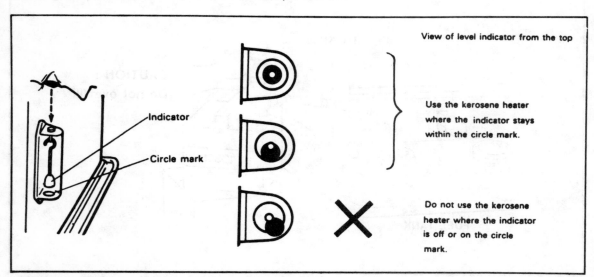

View of level indicator from the top

Use the kerosene heater where the indicator stays within the circle mark.

Do not use the kerosene heater where the indicator is off or on the circle mark.

Indicator

Circle mark

Fig. 3-12. Level indicator (courtesy Robeson Industries Corporation).

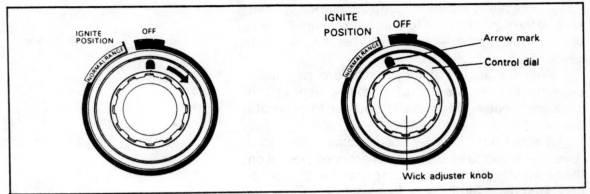

Fig. 3-13. Wick adjuster knob to "ON" (courtesy Robeson Industries Corporation).

Fig. 3-14. Wick adjuster knob to "IGNITE POSITION" (courtesy Robeson Industries Corporation).

weight that will automatically extinguish the heater should it be accidentally bumped or tipped over.

Fully turn the wick adjuster knob clockwise to raise the wick to the maximum height. When the wick is completely moved up, the arrow mark of the control dial will point to the "IGNITE POSITION" (Fig. 3-14).

To light the heater with the ignition device, simply push down the ignition knob. The burner cylinder tilts for ignition (Fig. 3-15).

If the ignition device doesn't work, check and replace the ignition coil or dead battery with a new one. If a match is to be used, open the guard and perform ignition as shown in Fig. 3-16. Never leave a burnt match on the burner basket or shield plate as incomplete burning or fire may occur.

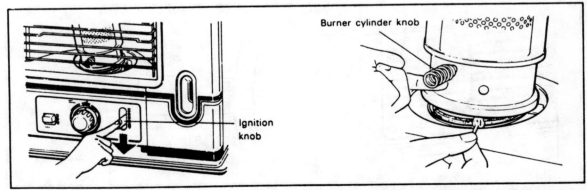

Fig. 3-15. Lighting with the ignition switch (courtesy Robeson Industries Corporation).

Fig. 3-16. Lighting with a match (courtesy Robeson Industries Corporation).

After igniting the wick, rotate the burner cylinder knob to the right and left a few times to ensure that the burner rests properly on the burner basket (Fig. 3-17).

Wait for about 10 minutes before reigniting. Lighting the wick soon after extinguishing the kerosene heater will cause the burner to generate heavy odor.

A slight odor may be noticed during the initial use. This is due to evaporation of anticorrosion oil on the new heater. This odor will be gone after the burner has been used for three or four hours.

Lowering of the ignition knob with excessive force may allow the ignition coil to be caught by the wick. Ignition may not occur. Raise the ignition knob slightly and lighting becomes easier. Raising it too much will result in ignition failure because the ignition coil is too far away from the wick (Fig. 3-18).

ADJUSTING THE FLAME

Adjusting the flame on your kerosene heater can improve efficiency and safety. Look at the wick adjuster knob on your kerosene heater. A typical one looks like that in Fig. 3-19. The flame becomes stronger when the wick adjuster knob is turned in the direction of "ON" (clockwise) and weaker when it is turned in the direction of "OFF" (counterclockwise).

Within 20 to 30 minutes after ignition, the flame may become higher with the hotter room temperature. Hold the burner cylinder knob and rotate the

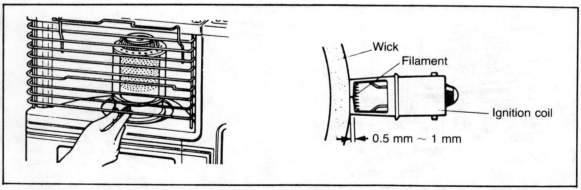

Fig. 3-17. Rotate the burner cylinder knob (courtesy Robeson Industries Corporation).

Fig. 3-18. Check the distance of the ignition coil (courtesy Robeson Industries Corporation).

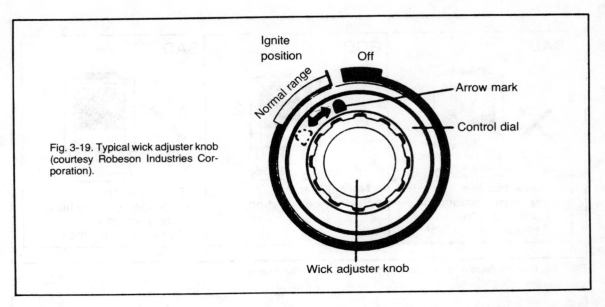

Ignite position

Off

Arrow mark

Normal range

Control dial

Fig. 3-19. Typical wick adjuster knob (courtesy Robeson Industries Corporation).

Wick adjuster knob

burner cylinder right and left gently to see if the flame becomes stable. If the flame stays higher, lower the wick slightly to obtain a normal flame (Fig. 3-20).

Regulating the dial out of the "NORMAL RANGE" will produce carbon on the wick and burner basket and result in improper combustion or in stiffening the lower and upper parts of the wick. To obtain a normal flame, keep the arrow mark of the control dial within the "NORMAL RANGE" (Fig. 3-19).

Some kerosene heaters have simpler wick controls (Fig. 3-21).

EXTINGUISHING YOUR HEATER

On most models, push down the extinguishing knob to extinguish the flame. If the wick is not lowered by the action, keep the extinguishing knob pushed down and turn the wick adjuster knob counterclockwise ("OFF" direction) to the limit. In about two to three minutes, lift up the burner cylinder knob to confirm extinguishment (Fig. 3-22).

When a large amount of carbon adheres to the wick, it may not lower completely even by pushing down the extinguishing knob. This could mean that the flame continues to burn. In this case, use the alternate method of shutting down and perform the inspection and maintenance steps described next.

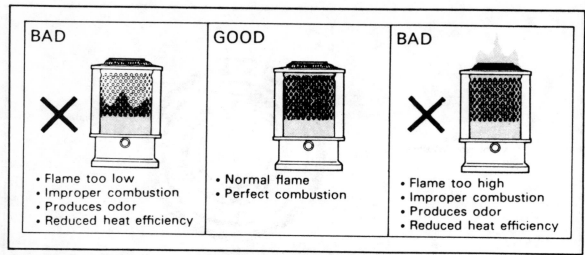

BAD	GOOD	BAD
• Flame too low • Improper combustion • Produces odor • Reduced heat efficiency	• Normal flame • Perfect combustion	• Flame too high • Improper combustion • Produces odor • Reduced heat efficiency

Fig. 3-20. Flame adjustments (courtesy Robeson Industries Corporation).

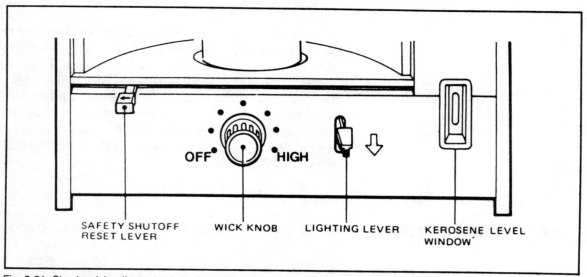

Fig. 3-21. Simple wick adjuster knob (courtesy Robeson Industries Corporation).

A daily check and maintenance routine will take only a few moments, but it can substantially extend the life of your kerosene heater (Figs. 3-23 through 3-27). Always keep your kerosene heater clean. The reflector, if any, should be free of dust, film, grease, oil, or other combustible substances. Using a dirty kerosene heater is dangerous and also shortens the heater's lift. Here is a simple six-step daily inspection.

DAILY MAINTENANCE

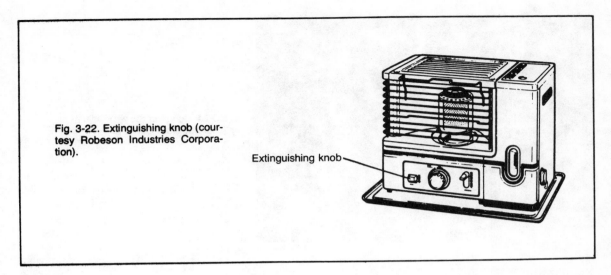

Fig. 3-22. Extinguishing knob (courtesy Robeson Industries Corporation).

Extinguishing knob

Step One

Inspect the cartridge tank or tank assembly. Are there any signs of a kerosene leak? Is there rust or foreign matter in the tank? Are there any dents or cracks in the tank? Is the tank's cap in good condition? Replace defective parts as needed.

Step Two

Inspect the burner basket and wick guide cylinder as often as possible to see if carbon has accumulated. If it has, remove it as outlined later in this chapter under "Wick Maintenance."

Fig. 3-23. Cleaning the wick adjuster (courtesy Kero-Sun, Inc.).

Fig. 3-24. Scouring the wick adjuster (courtesy Kero-Sun, Inc.).

Step Three

Inspect the ignition coil (Fig. 3-18). If the wick does not ignite or it is hard to ignite with battery ignition, check if the ignition coil filament is broken. A slight extension of the ignition coil filament can be corrected with a match stick. When the filament is extended out or broken, replace it with a new one. Insert the new ignition coil into the socket carefully so as not to damage the coil filament. Additional instruc-

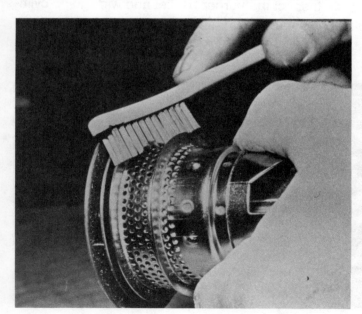

Fig. 3-25. Cleaning the flame spreader (courtesy Kero-Sun, Inc.).

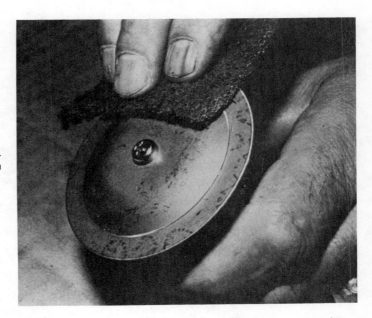

Fig. 3-26. Clean any other large surface areas that need attention (courtesy Kero-Sun, Inc).

tions on replacing the ignition coil follow. Remove batteries before replacing the ignition coil.

Step Four

In the event of low battery voltage (when the shape of the ignition coil filament is normal), the ignition coil filament will not heat red enough. Replace batteries in this case. They are normally mounted in the base or in the back of your kerosene heater.

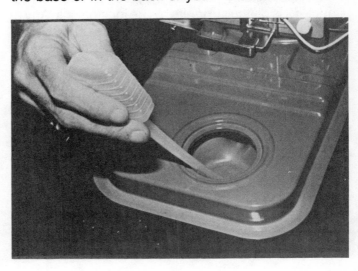

Fig. 3-27. Siphon out deposits in the sump tank (courtesy Kero-Sun, Inc.).

Step Five

Check the condition of the chimney coil for uniformity in spacing and breakage. If irregularity is evident, correct by hand. If the chimney coil must be replaced, follow instructions given later in this chapter.

Step Six

Check the condition of the wick. As noted earlier, make sure the proper quality of kerosene is used. Good kerosene is as clear as tap water and has no visible contaminants. Yellowish kerosene must be avoided. Poor kerosene will cause carbon to be deposited on the wick. This deposit restricts the flow of fuel and causes poor combustion. It will eventually completely block the flow of fuel.

WICK MAINTENANCE

The tops of fiberglass wicks accumulate carbon in the process of burning kerosene. Carbon must be removed to maintain peak efficiency in burning. You should feel for carbon hardness after the second or third tankful of fuel. A wick in good condition will feel soft while one that is carbonized will feel like a bristle brush. The top could be coated completely with hard black carbon in late stages.

Check for resistance when turning the wick adjuster knob. See if the chimney coil is heated red. Check if it is difficult to ignite by the ignition coil. Any of these troubles result from carbon on the upper part of the wick. To eliminate carbon, perform the following procedure.

A strong odor will be produced while removing carbon. It should be done outside on a windless day or inside with the windows open for ventilation.

In the case of the cartridge tank having only a small amount of kerosene, keep the wick burning without refilling with kerosene even when the tank becomes empty. If a substantial amount of kerosene is in the tank, lower the wick fully to extinguish the kerosene heater. Take the tank out of the heater after the flame is extinguished, wait about two minutes, then turn up the wick fully and reignite.

When the red heat of the chimney coil becomes faint, turn up the wick fully. Leave it there for about an hour or until it burns out.

The upper part of the wick will be rid of carbon and softened with this operation. If any parts are still left stiff, pinch them with small pliers to fracture the carbon into pieces. Supply a small amount of kerosene to the tank and repeat the above steps.

If you follow these steps for removing carbon with five to seven days after the first use of the kerosene heater, carbon accumulation to the wick will be reduced. Perform the steps when the wick becomes stiffened due to carbon buildup—about once a week during the heating season.

KEROSENE FUEL GUIDELINES

There may be no fuel anywhere that is called as many different names as is kerosene: lamp oil, stove oil, coal oil, or range oil. Sometimes even major oil companies call kerosene "No. 1 fuel oil." Many oil companies sell a kerosenelike product under that name with a higher sulfur content, higher carbon residue, and lower API (American Petroleum Institute) gravity. This combination will result in very inefficient fuel usage and, quite often, soot and smoke.

Use pure kerosene grade K-1 low sulfur. Never use gasoline, naphtha, or solvents such as alcohol, acetone, and petroleum distillates. They are hazardous and may cause fire, explosion, and serious injury. Refuel outdoors when the heater is cool to the touch.

Figure 3-28 illustrates the best way to check for good kerosene. Measure specific gravity by a hydrometer so you can see the difference of the kerosene. The specific gravity of individual petroleum products is different.

Specific Gravity

Gasoline	0.71-0.72
Diesel oil	0.82-0.85
Light oil	0.84-0.85
Jet oil	0.78
Kerosene	0.79-0.81

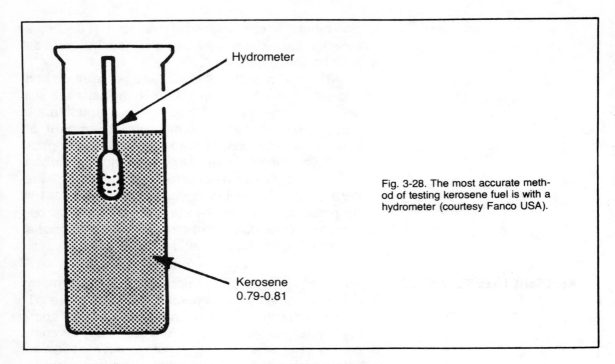

Fig. 3-28. The most accurate method of testing kerosene fuel is with a hydrometer (courtesy Fanco USA).

Hydrometer

Kerosene
0.79-0.81

Kerosene is *not* the same as diesel oil or jet oil. Good quality kerosene will meet the following criteria:

Reaction test	Neutral
Ignition point	104° F. minimum
95 percent distillate temperature	518° F. maximum
Sulfur content	0.015 percent maximum
Smoke point	0.9 inches minimum
Specific gravity	0.8

The best method of insuring an ample supply of good quality kerosene is to shop around. Talk with kerosene heater dealers, users, petroleum dealers, fire officials, and others. When you've found a reliable source, stay with it. Use only top quality K-1 low sulfur kerosene for the safety and health of yourself and your family.

Someday you may get some kerosene that is of a low quality or has deteriorated. Completely drain the

DEALING WITH BAD KEROSENE

kerosene from the fuel tank and pump it dry a few times. If *any* old kerosene is left in the tank, trouble may occur.

Kerosene may deteriorate in several ways. Kerosene that has been stored in a polyethylene container for several months will deteriorate. Some other types of containers allow kerosene to deteriorate. Kerosene subjected to direct sunlight for an extended period can go bad. Kerosene stored in an uncapped container can deteriorate.

Impurities can also contaminate kerosene and make it useless or even dangerous. Possible impurities include other oils such as machine or crude oil—even in a very small amount—water or dirt, rust, or some metals. The most common contamination is from storing kerosene in a container that has been used for other fuels. Always clean the interior of old containers by rinsing with clean pure kerosene several times before use. The same goes for siphon pumps.

Selecting an inferior grade of kerosene, either by choice or by carelessness, is false economy. Lower grade kerosene containing a higher amount of sulfur can reduce the life and efficiency of the costly wick and other components. Poor kerosene must be thrown out and good kerosene used to rinse out containers—a real waste.

More important, using a fuel other than water-free grade K-1 low sulfur kerosene will probably void the warranty on your heater. Most important, using incorrect, deteriorated, or impure fuel in your kerosene heater is simply unsafe. Through understanding kerosene, its properties, and how to select and use it in your kerosene heater, you can insure both comfort and safety.

Chapter 4

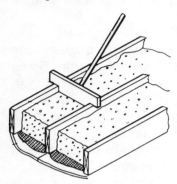

Increasing Kerosene Heater Efficiency

THE PORTABLE KEROSENE HEATER OFFERS EFFIcient space or zone heating in the most used portions of your home so that you can reduce the setting and the cost of your primary heating system. This savings can only be realized if your home has first been insulated properly to increase kerosene heater efficiency. By insulating your home you can reduce your heating needs, thereby reducing the size and the cost of your portable kerosene heater.

A key to cutting energy use in your home is thorough insulation of ceilings, walls, floors, ducts, and hot water pipes. Caulking, weather stripping, and installing storm windows and doors complete the job.

Because most homes were built before energy became a national concern, homes were seriously underinsulated if insulated at all. Heating and cooling systems were oversized in order to overcome structural deficiencies. Energy then was cheap and abundant.

Now homes are operated on expensive, questionably available energy. Homeowners have turned down the main heating system and used cost-efficient supplementary heating systems such as kerosene heaters. Even this is not always enough. Greater energy efficiency can be realized with proper weatherization, too.

HOME INSULATION BASICS

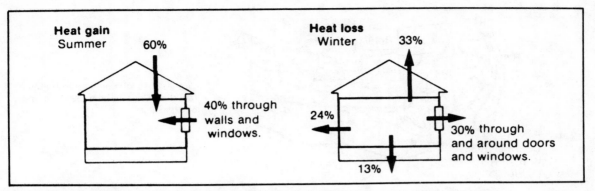

Fig. 4-1. Summer heat gain and winter heat loss (courtesy United States Department of Agriculture).

Heat passes back and forth through your house. Insulation slows this process and keeps warm and cool air where you want it when you need it.

Figure 4-1 shows that summer heat gain is mostly through the ceiling, walls, and windows. Winter heat loss occurs in the same places, but also significantly through the floor.

Additional heat gains and losses occur by infiltration through large and small cracks and holes. Air also comes into the home when doors are opened and closed as family members move in and out. In addition, opening a window slightly to allow for portable kerosene heater ventilation can also cause heat loss.

Air changes are needed in the home to maintain a healthful environment. The goal of good weatherization is simply to slow the process of unnecessary air changes and control the heating and cooling to an adequate, healthful level.

R VALUE Insulation needed in ceiling, walls, and floors depends on where your house is located. Climate zones have been drawn on a map of the United States to enable you to determine these requirements easily (Fig. 4-2).

The amount of insulation needed is expressed as an *R value* or R factor—an "R" followed by a number. The combination of numbers for ceilings (or attic floors), walls, and floors is different, and each zone is different. You can determine basic require-

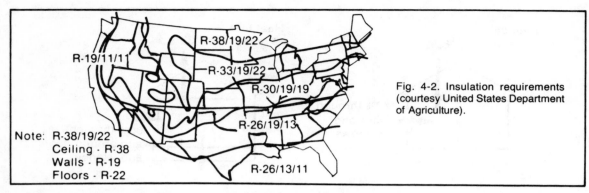

R-38/19/22

R-19/11/11

R-33/19/22

R-30/19/19

R-26/19/13

Note: R-38/19/22
Ceiling - R-38
Walls - R-19
Floors - R-22

R-26/13/11

Fig. 4-2. Insulation requirements (courtesy United States Department of Agriculture).

ments for your home by locating it on the map and simply reading the R value levels needed.

The R value of an insulating material is the most reliable indication of the job it can do for you when properly installed. This value indicates the material's ability to resist the flow of heat passing through it—the more resistance, the higher the R value, and the better job the material can do.

Some commonly used insulation materials, the forms in which they are manufactured, and the number of inches of thickness of each needed to achieve the R value required are listed in Table 4-1.

Besides using the climate zone map to determine your insulation levels, it's a good idea to consult an energy supplier in your area. Many utilities recommend somewhat higher levels and may, in turn, give lower rates for homes with higher levels of insulation.

Batts, blankets, loose fill, blown-in, and rigid boards are the main forms in which insulating mate-

INSULATION FORMS

R Value	Glass Fiber	Rock Wool	Glass Fiber	Rock Wool	Cellulose Fiber	Vermiculite	Perlite
R-11	3½	3	5	4	3	5	4
R-13	4	3½	6	4½	3½	6	5
R-19	6	5	8½	6½	5	9	7
R-22	7	6	10	7½	6	10½	8
R-26	8	7	12	9	7	12½	9½
R-30	9½	8	13½	10	8	14	11
R-33	10½	9	15	11	9	15½	12
R-38	12	10½	17	13	10	18	14

Table 4-1. Nominal R Values for Various Thicknesses of Insulation (courtesy United States Department of Agriculture).

(inches of thickness)

rials are manufactured. Actual material content of the form may vary.

A particular material or form chosen should be used only for the purpose and place for which it was intended. Inappropriate use, such as a fiberglass batt beneath a concrete slab, destroys the material's effectiveness (in this case by compacting it) and usually voids any manufacturer's warranty.

Let's examine each type of insulation (Figs. 4-3 and 4-4).

Batts of glass fiber or rock wool are used to insulate unfinished attic floors, unfinished attic rafters, the underside of floors, and open sidewalls. They are popular with the do-it-yourselfer, are easy to handle, and fit standard joist and stud spacings. Batts can be wasteful if they need to be cut to fit.

Blankets of glass fiber or rock wool are used to insulate unfinished attic floors, unfinished attic rafters, the underside of floors, and open sidewalls. They are also popular with the do-it-yourselfer, have less waste due to hand cutting of needed lengths, and they fit standard joists and stud spacings.

Foamed-in-place insulation, also known as expanded *urethane*, is used to insulate finished frame walls only and is installed only by contractors. It is

Batts — glass fiber, rock wool

Where they're used to insulate:

    unfinished attic floor
    unfinished attic rafters
    underside of floors
    open sidewalls

Blankets — glass fiber, rock wool

Where they're used to insulate:

    unfinished attic floor
    unfinished attic rafters
    underside of floors
    open sidewalls

Foamed-in-place — expanded urethane

Where it's used to insulate:

    — finished frame walls only

Fig. 4-3. Insulation types (courtesy United States Department of Agriculture).

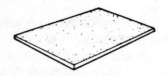

Rigid board—polystyrene (extruded), expanded urethane (preformed), glass fiber, polystrene (molded beads)

Where it's used to insulate:

    exterior wall sheathing
    floor slab perimeter
    basement masonry walls

Loose fill (blown-in) — glass fiber, rock wool, cellulose

Where it's used to insulate:

    unfinished attic floor
    finished attic floor
    finished frame walls
    underside of floors

Loose fill (poured-in) — glass fiber, rock wool cellulose, vermiculite, perlite

Where it's used to insulate:

    unfinished attic floor

Fig. 4-4. More insulation types (courtesy United States Department of Agriculture).

often used to add insulation within the walls of older homes.

Rigid board of extruded polystyrene, expanded urethane, glass fiber, or molded-bead polystyrene is often used to insulate exterior walls, floor slab perimeters, and basement masonry walls. Rigid board insulation offers a high insulating value for thickness. It will not compact easily and must be covered for fire safety.

Loose fill insulation of glass fiber, rock wool, or cellulose is blown in to finished and unfinished attic floors, finished frame walls, and the underside of floors. It is easy to use around obstructions and hard-to-reach spaces. Loose fill can also be poured in to insulate unfinished attic floors.

Consider an insulation material's flammability, ability to take up or hold water, and stability. Does it stay where you put it? Does it pack easily? Does it shrink?

VAPOR BARRIERS

Batts and blankets come with and without *vapor barriers* or retarders attached in the form of kraft paper that is often sprayed with asphalt or layered with polyethylene or foil. Rigid boards are impermeable, but loose fill often needs a vapor retarder added.

An acceptable effective moisture retarding material is any material with a perm rating less than 1. The rating is a measure of a material's permeability or characteristic of allowing moisture vapor to pass through.

Aluminum foil has a perm rating of 0. If perfectly installed, it would allow no moisture vapor passage. Other materials in use have some permeability but are effective retarders.

Moisture vapor is generated in a surprisingly large quantity in a home simply from breathing, bathing, clothes and dish washing, and preparing food. Kerosene heaters add moisture to the air.

Moisture vapor passes *through* the structure of a home and condenses or forms water drops when it hits a cold surface (Figs. 4-5 and 4-6). To stop this from happening, especially in walls, vapor retarders are installed at the same time insulation is put in place. Vapor retarders are placed *toward* the winter heated space.

When you are seated in the living room, a vapor retarder will be beneath your feet just below the sub-flooring, in the walls, just beneath the gypsum board or paneling, and most often overhead, just above the ceiling gypsum board or tile.

VENTILATION

Besides vapor retarders, adequate and carefully controlled ventilation must be provided, especially when burning unvented kerosene heaters. Crawl spaces, as a minimum, need 1 square foot of clear vent area

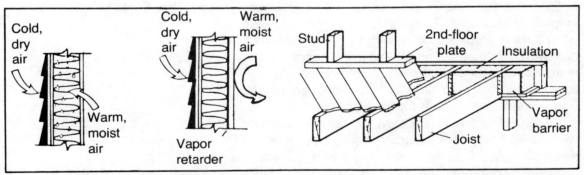

Fig. 4-5. Vapor in walls (courtesy United States Department of Agriculture).

Fig. 4-6. Vapor barrier in floor (courtesy United States Department of Agriculture).

Fig. 4-7. Attic ventilation methods (courtesy United States Department of Agriculture).

for every 150 square feet of first floor space. Attics need about half that amount in a combination of low eave vents and high ridge area vents (Figs. 4-7 through 4-9).

Exhaust fans inside your home and vented to the outside—operated in baths, laundries, and kitchens at times of high moisture vapor production—aid the internal environment when excessive moisture conditions prevail.

Insulation that you install may pay for itself in energy cost savings in three or five years or certainly in about a 9 percent higher resale value of your home if you plan to move soon. Income tax credits help, too, so there's no good argument for not saving energy and money by insulating. A well-insulated home will be more adequately heated with a smaller portable heater than a poorly insulated home.

Let's look at the steps you should take to adequately insulate your home. The first thing to do is decide what form of insulation you want to use. This will be dictated by its use, whether it is owner- or contractor-installed, cost and local availability. Review the different types of insulation outlined and select the one(s) most appropriate for your home.

Figure the square footage for ceilings, exterior walls, and first floor (including floors over any unheated area such as carports). Joists and studs take up about 10 percent of the area, so you will be insulating about 90 percent of the area.

GETTING STARTED

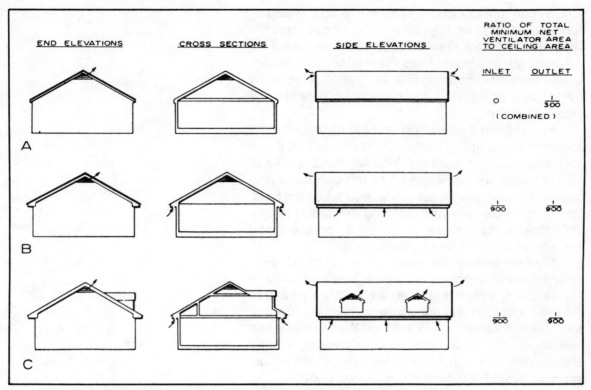

Fig. 4-8. Ventilating gable roofs (courtesy United States Department of Agriculture).

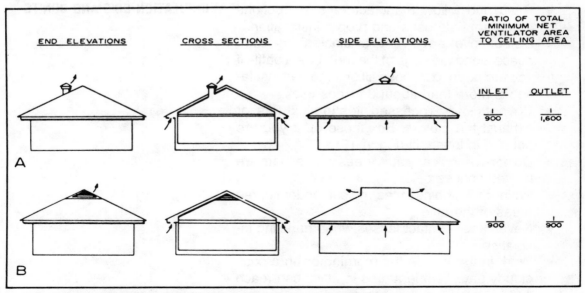

Fig. 4-9. Ventilating hip roofs (courtesy United States Department of Agriculture).

Get an amount and cost estimate on the insulation you will need for the job. Also, get an agreement about unused amounts if you buy all at one time. This will minimize waste and keep the total cost down. Get an estimate for increasing attic ventilation or adding a crawl space covering of 4 to 6-mil polyethylene if not aleady in place.

You will need a sharp knife or long-bladed scissors and a rigid straightedge, such as a board, if you decide on batts or blankets. You will need a rake to evenly place poured loose fill or to push batts and blankets into narrow spaces. Bamboo rakes are best. Obtain a staple gun for fastening insulation in any unfinished walls. You will also need a portable light if the attic or crawl space is not lighted.

Get some ¾-inch plywood to support your weight between joists. If you walk between joists in the attic, you will probably go through the ceiling to the rooms below. Get some insulation supporters or wire for insulation installed in floors. Wear clothes that fully cover you, including gloves. Cover your nose and mouth with a handkerchief, gauze, or a dust mask to avoid breathing dust or small fibers.

INSULATION DO'S AND DON'TS

- Read and follow manufacturer's instructions for appropriate uses and proper installation.
- Do not cover eave vents or block air passage space along the edge of the roof. Use a baffle if you are pouring in loose fill or if the batt insulation is more than about 6 inches thick.
- Do not cover recessed lighting fixtures or exhaust fan motors. Box these off if you are pouring in loose fill (Fig. 4-10).
- Do not overlook any attic areas where there are heated spaces below.
- Push insulation as far as you can under floored areas of the attic.
- Never wear contact lenses when handling insulation.
- Work in the attic in the morning or on a cool, cloudy day. Temperatures in attics can reach 140° F.

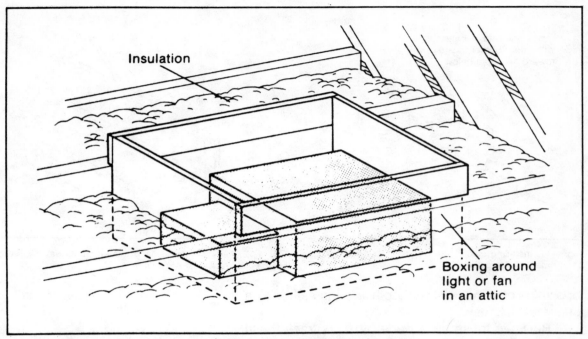

Insulation

Boxing around
light or fan
in an attic

Fig. 4-10. Boxing insulation (courtesy United States Department of Agriculture).

- Watch out for nails sticking through roof sheathing or subflooring.
- Take a cold shower when you finish. Cold water closes pores and washes off insulation particles.

UNINSULATED CEILINGS

Here's how to install insulation in uninsulated ceilings. First, unroll blankets or place batts between attic joists (Fig. 4-11). When you encounter wiring, slip the material under the wires. When you encounter bracing, cut the material and place it tightly above and below the bracing. Be sure the vapor retarder, if used, is placed downward toward the heated living area of the house.

Start from the edges of the attic and work toward the center. Cutting most likely will occur where there is more headroom.

If you have chosen to pour in loose fill insulation, simply pour the insulation from the bag into the space between the joists to the thickness needed for the proper R value (Figs. 4-12 and 4-13). Use the bam-

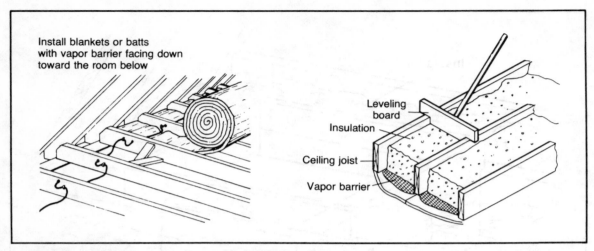

Install blankets or batts with vapor barrier facing down toward the room below

Leveling board
Insulation
Ceiling joist
Vapor barrier

Fig. 4-11. Installing attic insulation (courtesy United States Department of Agriculture).

Fig. 4-12. Leveling loose fill insulation (courtesy United States Department of Agriculture).

boo rake or a board to smooth the insulation to a uniform thickness.

Blown-in loose fill is done by a contractor with special pneumatic machinery, with much the same results as when loose fill is poured in.

Vapor retarders need to be added when pouring or blowing in loose fill. Plastic sheeting may be used, or the interior surface of the ceiling may be painted with special vapor-retarding paint or wallpapered with plastic-coated wallpaper.

Use batts, blankets, or loose fill to aid insulation to get R values you need. Buy batts and blankets without attached vapor retarders. Otherwise, cut the retardant material about every foot with a knife or tear it off. An added vapor retarder on top of existing insulation would trap moisture vapor.

These types of insulation may be used in any combination. First, determine what you have in the way of R values and vapor retarders. Then simply add a vapor retarder to the ceilings below if you need one and add insulation to the attic floor to achieve the full R value level you need.

PARTIALLY INSULATED CEILINGS

INSULATING WALLS

Determine how much insulation you have, if any. Turn off the electricity and remove the cover of a convenience outlet or light switch plate located on an outside wall to check.

If there is some insulation in the wall cavity, do not plan to add more. If there is none, the walls can be insulated by a contractor with special equipment to blow in or foam in place the material you select.

Holes about 2 inches in diameter are drilled in each wall cavity. The insulation is put in through the holes. The holes are then plugged with a precut wooden plug. Holes are drilled from the outside where possible.

A vapor retarder will have to be added to the inside surface of all outside walls in the form of two coats of oil-based paint made especially for this purpose, or plastic- or aluminum foil-coated wallpaper.

The job is simpler and less costly when wall cavities are exposed, as when a house is under construction or complete remodeling is being done. Often studs are exposed in unfinished garages. Blankets and batts with attached vapor retarders are used. The material is stapled in place.

If the vapor retarder is kraft paper, simply staple the flange of the material to the edge or face of the stud facing you inside the house (Figs. 4-14 and 4-15). Two overlapping flanges can be stapled at once, spaced every 3 to 5 inches apart. Staple only

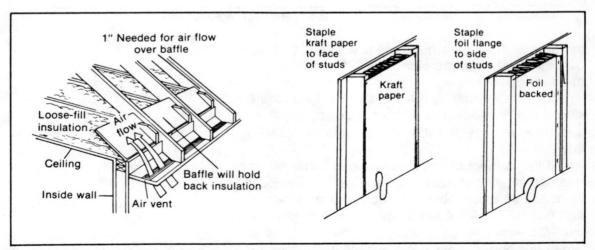

Fig. 4-13. Install baffles to allow air circulation (courtesy United States Department of Agriculture).

Fig. 4-14. Vapor retarding insulation (courtesy United States Department of Agriculture).

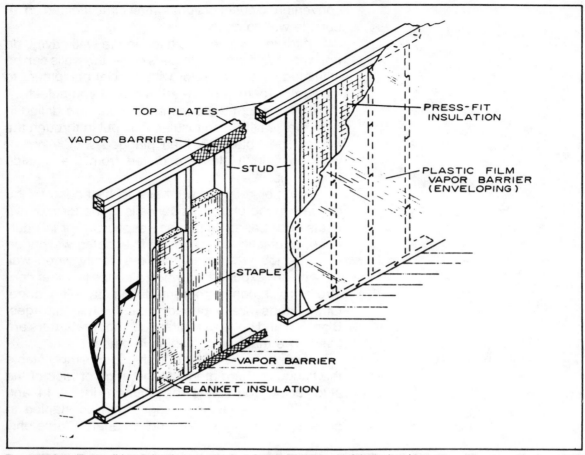

Fig. 4-15. Installing wall insulation (courtesy United States Department of Agriculture).

the flange. Take care not to allow insulation to lap over the stud face and create a bulge that will show on the finished wall later.

If the vapor retarder is foil, staple the flange to the side of the stud facing you. This creates a ¾-inch air space between the foil and the finished wall installed later.

If the batts or blankets have no attached vapor retarder, press them into the wall cavity. Compact them as little as possible. Fill cracks around windows and doors (Fig. 4-16) and staple 6-mil polyethylene over the entire wall, including windows and doors. When finished, cut away the polyethylene from the openings. Cracks around openings (windows and

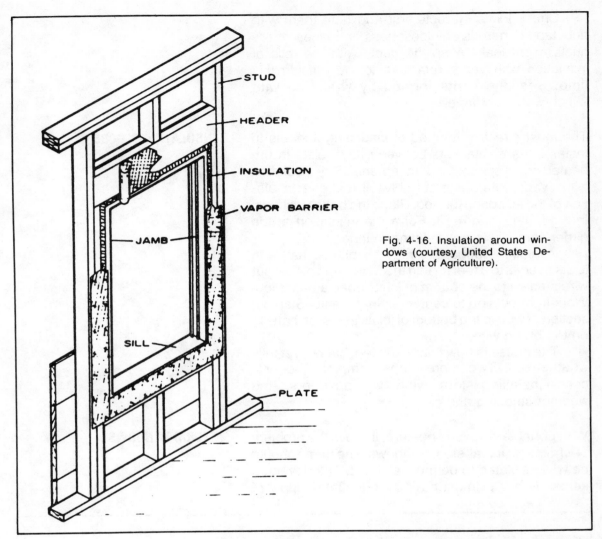

STUD

HEADER

INSULATION

VAPOR BARRIER

JAMB

SILL

PLATE

Fig. 4-16. Insulation around windows (courtesy United States Department of Agriculture).

doors) should be stuffed and covered with polyethylene when using the materials with attached vapor retarders.

Rigid insulating boards achieve higher R values. They are installed on the outer side of the wall cavity in new and full remodeling jobs.

INSULATING DUCTS

Ducts passing through unconditioned space (not heated or cooled) must be insulated with special insulation manufactured for this purpose. Use the type that is 2 inches thick with a vapor retarder attached.

Check joints in ducts first and tape them with duct tape if there is any looseness or spaces where ducts might leak. Wrap the ducts with the rolls of insulation with vapor retarders to the outside this time. Seal the joints formed by wrapping with 2-inch-wide duct tape.

INSULATING FLOORS

The most effective method of insulating floors is to install batts or blankets between floor joists in unheated crawl spaces and basements. Buy insulation with a vapor retarder, preferably foil, for better insulation of the air space formed. Place the batts as shown in Figs. 4-17 and 4-18. Form the insulation at the girder carefully in a well-fitting manner.

Insulation supporters may be placed between joists every 2 to 3 feet to hold the material in place, but wire stapled to the bottom of joists does a better job. Work from outside to center, as in the attic. Staple a section of wire to the bottom of joists and slide batts in on top of the wire.

The house is now fully insulated, but not yet fully weatherized. Two more steps complete the process—installing storm windows and doors, and weather stripping and caulking.

STORM WINDOWS

With portable kerosene heaters, there is the problem of whether to install storm windows. The home should be well-insulated to be most efficient. The unvented kerosene heater must have an adequate supply of

Fold

Fig. 4-17. Fold the blanket end for installation (courtesy United States Department of Agriculture).

96

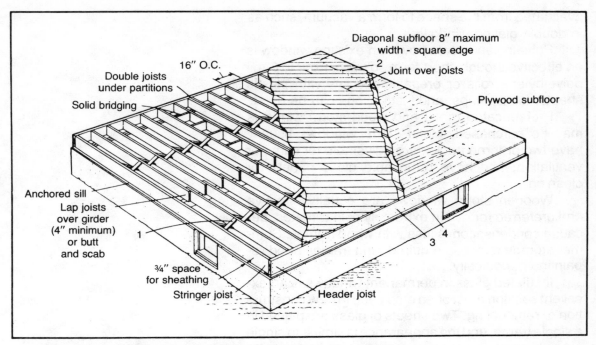

Fig. 4-18. Typical floor framing (courtesy United States Department of Agriculture).

fresh air to work efficiently and safely. An unvented kerosene heater in a tightly insulated room may burn up the oxygen supply. Occupants of the room may suffocate. An unvented heater can also trap harmful gases within the house.

The solution depends on how much ventilation naturally occurs in your home. Unvented natural gas stoves have been used in homes for years with minimal problems due to the leakage of air through walls, floors, and ceilings. The tighter the home, the less air leakage.

Some homeowners overcome this dilemma by venting rooms to the interior of the home. The ventilation system replaces air used by the kerosene heater. The influx of air must be adequate to insure the safety of occupants.

Several methods can be used to form an insulating dead air space of ½-inch to 4½-inch thickness at windows. A minimum ½-inch air space between parallel surfaces is required to provide insulating value. Less than ½ inch is acceptable only if air is evacuated from the space to form a vacuum, such as in double glazed windows.

Polyethylene sealed over an existing window is an effective though short-term solution. Four to 6-mil polyethylene rolls or prepackaged kits with plastic sheets, tacks, strips, and instructions can be used.

Prefabricated metal and wood storm windows may be purchased. Metal storm windows usually have two or three tracks, are adjustable for summer ventilation, and have removable glass panels for cleaning.

Wooden storm windows are equally effective and preferred for use in extremely cold climates, because condensation is less with wood frames. Summer storage space is required, and frames need repainting periodically.

Insulated glass in permanent windows is an excellent solution most often employed in new construction or remodeling. Two sheets of glass are placed in a single frame, and the appearance is similar to single glazed windows. There are only two surfaces to clean. Breakage replacement costs are doubled, however.

Many homeowners are choosing to replace the single pane windows in their homes with a new double or triple glazed unit. This is a good solution. Replacement windows come in both wood and metal. Some wooden ones may be covered with vinyl for maintenance purposes. A metal replacement window should be used only if a thermal break has been built into the frame, and thermal or insulated glass is provided. Without a thermal break, heat will be rapidly conducted out of the house by the metal.

STORM DOORS

Storm doors are usually prefabricated metal, though wooden ones are sometimes used with equal effectiveness. Rigid frames, tempered safety glass or rigid nonbreakable plastic, automatic closing devices, and strong safety springs are important considerations. Convertible screen/storm doors are available and popular.

Value will be added to your home with the installation of all types except plastic sheeting. Strength of frames, good quality, warranties, and repair service are important considerations. If you don't do the entire job at one time, do the side of the house facing the prevailing winter winds first, the north side second, and the south side last.

WEATHER STRIPPING

Weather stripping comes in several forms and materials. Some are designed for use in one place, and some are more durable than others. Let's look at the different types of weather stripping, their installation, and use.

Self-adhesive foam type is a resilient sponge rubber or vinyl on paper or vinyl backing ⅜- to ¾-inch wide. It tends to deteriorate when exposed to weather and may last only one season. Apply foam tape to dry, clean surfaces at room temperature by pressing it in place on door and window jambs, stops, or sashes.

Felt or *aluminum and felt* is not the best weather stripping as it tears easily during use and is ineffective when wet. Its advantage is low cost. To install, staple to wood or glue to metal stops, sills, and sashes.

Vinyl weather stripping is durable. Apply it by simply tacking, stapling, screwing, or gluing the flange of the tube-shaped strip to surfaces.

Neoprene-coated sponge rubber is easy to install and is more durable than uncoated material. It, too, is tacked or stapled to surfaces.

Bronze metal weather stripping is durable and easy to install, but it is not affected by moisture and temperature. It is tacked to door and casement window jambs.

Caulking cords are easy to apply, pliable, durable, and not affected by moisture. They can be purchased in strips and pressed into place on any surface.

Fiberglass strip weather stripping is durable and comes in various sizes with waterproof tape. It seals larger cracks such as those around garage doors and may be wrapped around pipes for insulation.

Waterproof tape is not affected by moisture and is used to seal cracks. Apply half on the window sash and half on stops.

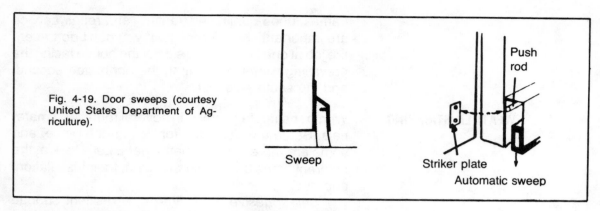

Fig. 4-19. Door sweeps (courtesy United States Department of Agriculture).

Sweep

Push rod

Striker plate

Automatic sweep

Air conditioner weather strip is low in cost and easy to install. The rectangular polyfoam strips are used for sealing around window-mounted units and window sashes.

Magnetic vinyl is a durable weather stripping used around steel doors.

Door bottoms sometimes are fitted with as large as a ¼-inch crack left at the bottom. This is equivalent to a 9-inch hole through your wall.

A brass-plated strip fastened to felt or vinyl may be attached to the inside bottom of the door. An even threshold is required, and level application is a little tricky to achieve. Other types of sweeps, even an automatically-operated sweep, are available (Figs. 4-19 and 4-20).

Thresholds may be improved to seal cracks. Replaceable vinyl bulb-shaped gaskets mounted in

THRESHOLDS

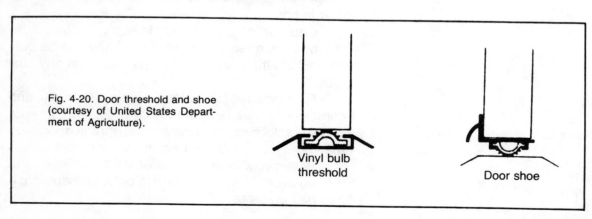

Fig. 4-20. Door threshold and shoe (courtesy of United States Department of Agriculture).

Vinyl bulb threshold

Door shoe

metal are effective when properly maintained. Combinations of door bottoms and thresholds are effective and longer wearing.

For additional information on insulating and weather stripping doors and windows, read my book, *Doors, Windows and Skylights* (TAB book no. 1578).

CAULKING

For a relatively small cost and time investment, large savings are produced when caulking is used. Caulking comes in toothpaste-size tubes and even 5-gallon buckets. A caulking gun with a tube that fits it usually is best. The tubes have directions for use and suggested places where they work best.

There are many products and prices. The less expensive may last only three to five years while the more expensive may last as long as 30 years. Some places you should check around your house for cracks and seams are: joints between door frames and siding, windowsills and siding, window frame and siding, window drip cap and siding, inside corners of a house formed by siding, joints where two things come together such as where steps and porches join the main part of the house, joints where the chimney and siding come together, around chimney and vent pipe flashings, and places where pipes, wires, and vents pass through exterior walls. Caulk hairline cracks and larger ones that you find.

SMART WEATHERIZING

Complete weatherization of a home may be done gradually rather than all in one big effort. Except for insulating finished walls, this is a do-it-yourself job for a homeowner in most cases.

The most effective steps are listed first with the less effective ones toward the end. This list is a general one that does not hold true for each home, because heat loss depends upon climate (Fig. 4-21), construction of the house, the shape, wall area, window and door area, etc.

First   Insulate the ceiling.
           Weather strip and caulk.

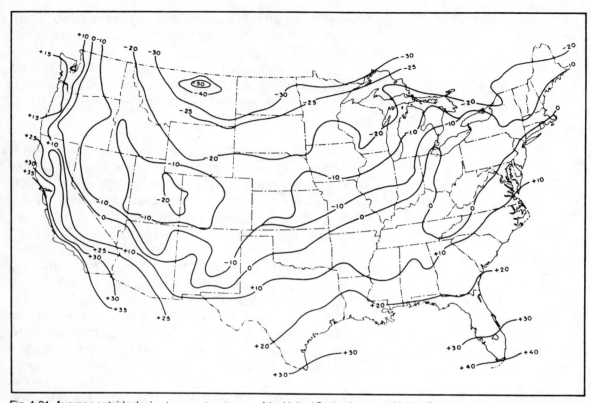

Fig. 4-21. Average outside design temperature zones of the United States (courtesy United States Department of Agriculture).

| Second | Install storm windows and doors. Insulate walls. |
| Third | Insulate floors. |

A completely weatherized house will use up to half as much fuel as it used when not weatherized. Combined with the savings of spot heating with a portable kerosene heater, your annual heating bill can be reduced drastically. There is a continued heat loss from completely weatherized homes. You have simply cut the rate of loss. Infiltration must continue to replace oxygen burned up by your kerosene heater. A healthy and cost-efficient balance can be obtained with appropriate weatherization and correct sizing of your portable kerosene heater.

Chapter 5

Maintaining Your Kerosene Heater

PORTABLE KEROSENE HEATERS ARE AN EFFICIENT and economical alternative energy source. They offer greater fuel efficiency than nearly all other heat sources and can be operated as zone heaters that reduce heating demands on your primary system.

Portable kerosene heaters depend on proper maintenance to insure continued heating efficiency and economy. Unlike many gadgets, kerosene heaters can be maintained and repaired by the owner. Proper kerosene heater maintenance can also improve the safety of your supplementary heating system and extend its life.

DAILY CHECK AND MAINTENANCE

In Chapter 3 you read about the simple daily maintenance checks you can make on your portable kerosene heater: inspection of tank, burner, ignition coil, batteries, and wick. Let's look at wick replacement in radiant and convection heaters—the primary maintenance item—then consider replacing other parts.

CLEANING THE WICK

You learned about the wick and how it works in Chapter 2. The wick used in portable kerosene heaters is a specially developed fiberglass wool or cotton type (Fig 5-1). The wick will not burn without fuel like conventional cotton wicks, but it needs proper attention to get long life.

103

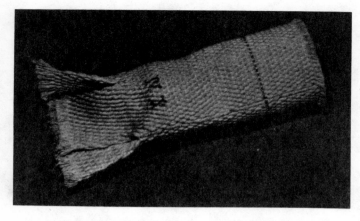

Fig. 5-1. The kerosene heater wick
(courtesy Kero-Sun, Inc.).

A kerosene heater wick should be burned off when the second or third tankful of kerosene has been used, when the wick control knob becomes hard to turn, when the radiator metal does not get red hot although the wick is turned up all the way, when the wick is hard to ignite, or when an unpleasant odor is produced during use.

Here's how to burn off a wick. A wick in good condition will feel soft. A wick that is carbonized will feel like a bristle brush and, in the last stages, the top might even be coated completely with hard black carbon. Use the purest kerosene available in your heater to slow this process—K-1 (low sulfur).

To restore a wick to its original softness, remove all the fuel from the tank or wait until the tank is empty. In a cartridge type heater you can simply remove the fuel cartridge. The remaining fuel in the reservoir will be used in the burn-off process.

Turn the wick to the highest position. Light it with a match if you cannot light it automatically. Be sure the chimney is properly centered. Let the wick burn until the fire is completely out.

If you feel that some carbon still remains, merely raise the chimney and relight with a match. Let it burn again until out. This burn-off should be done in a garage, on a porch, or in a place where odors will not be offensive.

Burning a wick dry does no harm to the wick itself. Allow the wick to cool, then remove any remaining ash with an old toothbrush (Figs. 5-2 through 5-5).

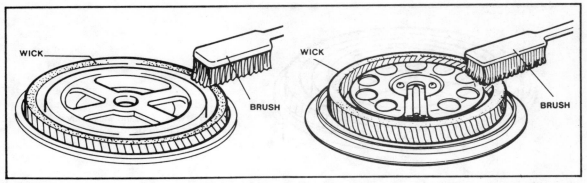

Fig. 5-2. Cleaning the radiant heater wick (courtesy Robeson Industries Corporation).

Fig. 5-3. Cleaning the convection heater wick (courtesy Robeson Industries Corporation).

You may vacuum away any residue or soot remaining in the burner area.

Using other than pure, clean, water-free grade K-1 kerosene (low sulfur) will damage the wick and probably void the warranty on your kerosene heater.

RADIANT HEATER WICK REPLACEMENT

The following steps and illustrations will show you how to replace the wick in a radiant type kerosene heater with an integral or built-in fuel tank.

Be sure that the top of the wick is even and level (Fig. 5-6). Wait approximately 20 minutes before lighting the heater the first time to let the wick draw up sufficient fuel. Don't cut, pull, or soil the new wick. Pulling the wick causes damage and results in poor combustion.

After several seasons or unusually heavy usage, the wick may have to be replaced. This procedure

Fig. 5-4. Cleaning the old wick (courtesy Kero-Sun, Inc.).

Fig. 5-5. The top of the wick must be free of carbon to operate properly (courtesy Kero-Sun, Inc.).

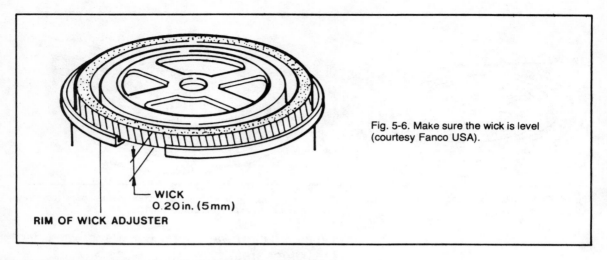

WICK
0.20 in. (5mm)

RIM OF WICK ADJUSTER

Fig. 5-6. Make sure the wick is level (courtesy Fanco USA).

should be performed on a completely cool heater after all the kerosene in the tank has been burned off. Allow the automatic shutoff to snap shut by jarring the heater body. Remove the batteries to prevent possible burns. Open the front grille and remove the chimney. Remove the safety shutoff reset lever by pulling it straight out of the cabinet. This lever is retained by means of a snap lock.

Remove the cabinet by unscrewing thumbscrews from the back and sides of the cabinet. Slide the electrical wire connectors off the battery case. Remove the screws at the side of the extinguisher assembly and lift it out. Remove the wick adjuster mechanism by loosening the wing nuts until you can turn the retainers that hold it in place.

To remove the wick from the adjuster, fold it and slide it off the rubber packing securing it (Fig. 5-7).

Here's how to install a new wick in your radiant heater with integral tank. Refer to Fig. 5-8. After turning the wick adjuster counterclockwise as far as it will go, fold the new wick and slide it into the adjuster. The red line on the outside of the wick should match the bottom edge of the adjuster. Turn the wick adjuster clockwise as far as it will go. Check the height of the wick, then press it against the teeth inside the adjuster to obtain a firm grip. Slide the rubber packing over the wick and allow the tails of the wick to drop down. Replace the wick and adjuster mechanism in the fuel tank. Make sure the wick fits evenly in place. Position the adjuster knob to the front of the heater. Tighten the wing nuts onto retainers. Turn the knob clockwise and counterclockwise a few times to make sure the mechanism is functioning smoothly. Recheck the height of the wick. If it has changed, readjust it. Any ragged edges that appear at the top of the wick should be trimmed with scissors.

Replace the cabinet on the fuel tank. Insert and tighten the thumbscrews that hold it in place.

The reset lever on the automatic shutoff is reinstalled by pushing it into its slot. The lever will snap into position. Slide the lever fully to the left to

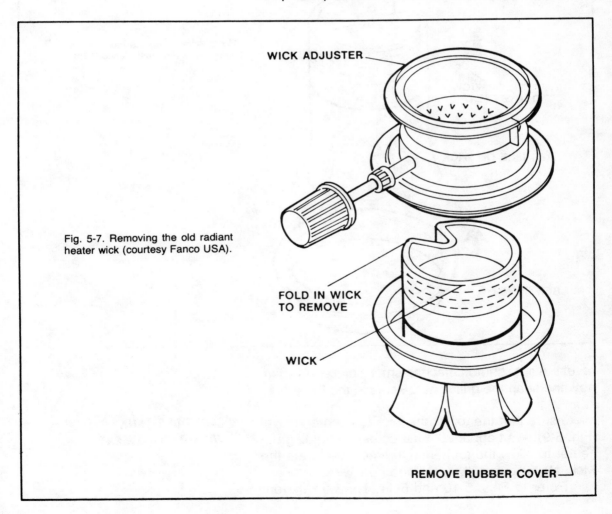

WICK ADJUSTER

Fig. 5-7. Removing the old radiant heater wick (courtesy Fanco USA).

FOLD IN WICK TO REMOVE

WICK

REMOVE RUBBER COVER

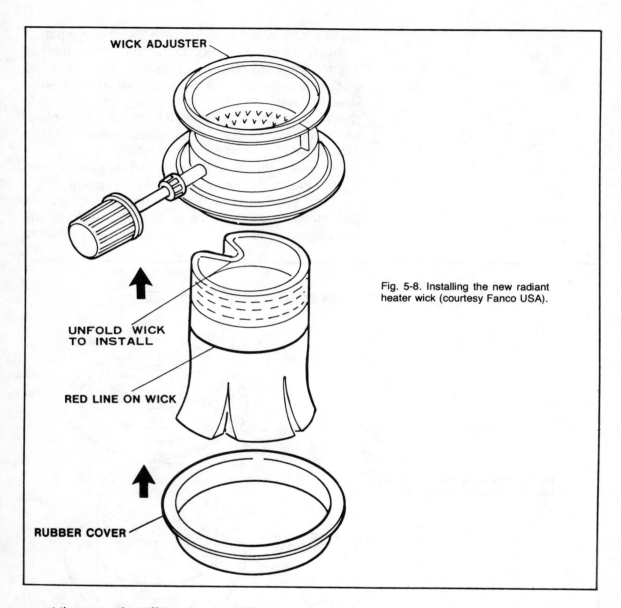

WICK ADJUSTER

UNFOLD WICK TO INSTALL

RED LINE ON WICK

RUBBER COVER

Fig. 5-8. Installing the new radiant heater wick (courtesy Fanco USA).

reset the auto shutoff mechanism. Replace the chimney and snap the grille shut. Replace the batteries.

Make sure that the top of the wick is even and level (Fig. 5-9). Wait about 20 minutes before lighting the heater to allow the fuel enough time to saturate the wick. Never cut, pull, or soil the new wick.

Refer to Figs. 5-10 and 5-11. The wick in your

CARTRIDGE TANK
WICK REPLACEMENT

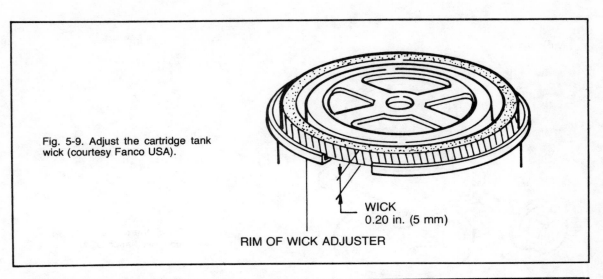

Fig. 5-9. Adjust the cartridge tank wick (courtesy Fanco USA).

WICK
0.20 in. (5 mm)

RIM OF WICK ADJUSTER

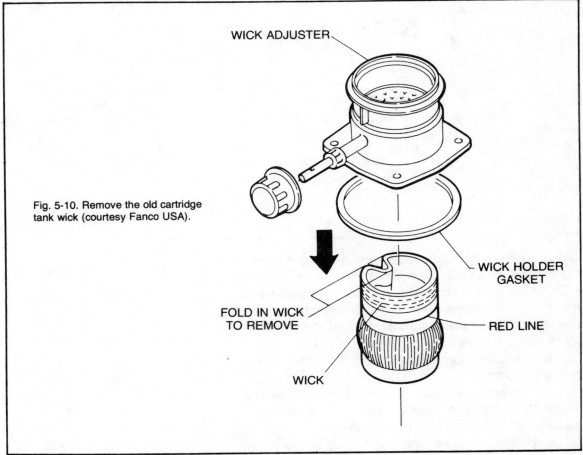

Fig. 5-10. Remove the old cartridge tank wick (courtesy Fanco USA).

WICK ADJUSTER

WICK HOLDER GASKET

FOLD IN WICK TO REMOVE

RED LINE

WICK

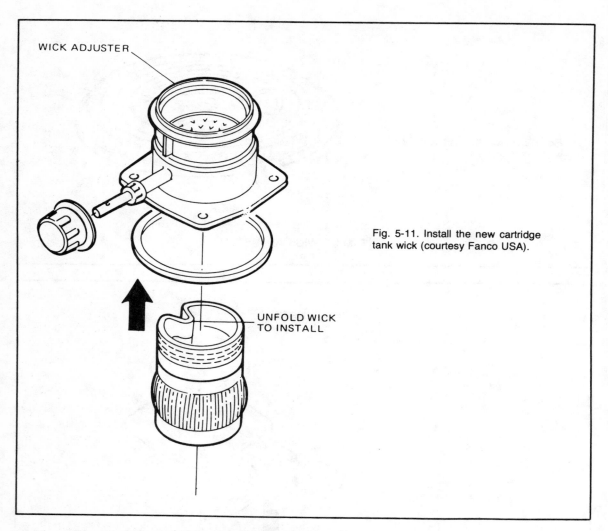

WICK ADJUSTER

UNFOLD WICK
TO INSTALL

Fig. 5-11. Install the new cartridge tank wick (courtesy Fanco USA).

heater will have to be replaced eventually due to use or damage. This should be done only on a completely cool heater after the cartridge has been removed and the remaining kerosene in the heater has been burned off. Allow the automatic shutoff to snap shut by jarring the heater body. Remove the batteries to prevent possible burns. Open the front grille and remove the chimney.

Remove the safety shutoff reset lever by pulling it straight out of the cabinet. Remove the wick adjuster knob by loosening the screws that hold it in place. Remove the ignition lever knob by carefully prying it with a regular screwdriver. Remove the cabinet by unscrewing the thumbscrews at the back and sides of the cabinet. Slide the electrical wire connectors off the battery case. Remove the screws at the side of the extinguisher assembly and lift it out.

Remove the wick adjuster mechanism by unscrewing the wing nuts that hold it in place. To remove the wick from the adjuster, fold it and slide it out.

Here's how to install a new wick in your cartridge tank radiant heater. After turning the wick adjuster counterclockwise as far as it will go, fold the new wick and slide it into the adjuster (Fig. 5-11). The red line on the outside of the wick should match the bottom edge of the adjuster. Turn the wick adjuster clockwise as far as it will go. Check the height of the wick as in Fig. 5-9. Press it against the teeth inside the adjuster to obtain a firm grip. Check to see that the gasket is still in place on the burner tank.

Replace the wick and adjuster mechanism in the burner tank. Make sure the wick fits evenly in place. Position the adjuster shaft to the front of the heater. Reinstall and tighten the wing nuts. Turn the knob clockwise and counterclockwise a few times to make sure the mechanism is functioning smoothly. Recheck the height of the wick. If it has changed, readjust. Any ragged edges that appear at the top of the wick should be trimmed with scissors. Replace the automatic extinguisher assembly with the screws and refasten the electrical wire connectors to the battery case.

Replace the cabinet. Replace the fuel cartridge to align the cabinet. Insert and tighten the screws at the back and sides of the cabinet that hold the cabinet in place. Add the adjuster knob and tighten the screw.

The reset lever of the automatic safety shutoff is reinstalled by pushing it into its slot. The lever will snap into position. Slide the lever fully to the left to reset the auto shutoff mechanism. The ignition lever knob is simply pushed onto its shaft. Replace the chimney and snap the grille shut.

Your radiant type kerosene heater may be slightly different in design than the model described. Nearly all such heaters work on the same principles with the same basic components and parts. You may have to modify instructions to fit your heater, but don't take shortcuts. Refer to your model's owner's manual for more specific instructions.

See Fig. 5-12. This illustrates how the wick in a convection heater will look and how much it should be adjusted. Be sure that the top of the wick is even and level. As with earlier models, wait about 20 minutes before lighting the heater for the first time so that the wick draws up sufficient fuel.

When replacing a wick in a convection heater, make sure that the heater is completely cool, the fuel tank is empty, and that the remainder of the kerosene in the heater has been burned off as discussed earlier. Jar the heater body to allow the automatic shutoff to snap shut. Remove the batteries. Remove the top plate by spreading the wire handle that holds it in place. Lift the burner unit out of the heater. Remove the outer cover by removing the screws that hold it in place. After taking off the wick adjuster knob (and safety reset knob on some models), lift off the body base. Remove the securing screws and lift off the shelter plate. Remove the wing nuts securing the wick

CONVECTION HEATER WICK REPLACEMENT

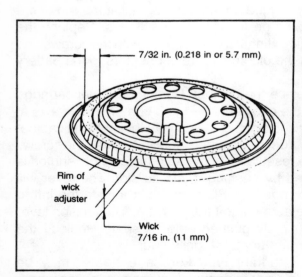

7/32 in. (0.218 in or 5.7 mm)

Rim of wick adjuster

Wick 7/16 in. (11 mm)

Fig. 5-12. Adjust the convection heater wick (courtesy Fanco USA).

Fig. 5-13. Removing the wick adjuster (courtesy Kero-Sun, Inc.).

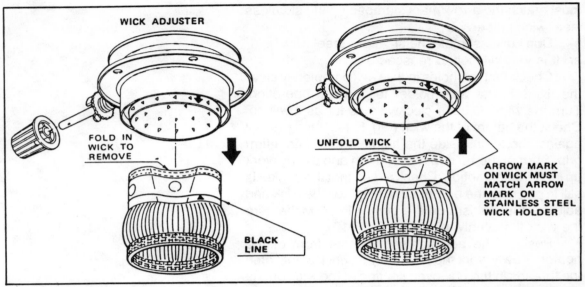

WICK ADJUSTER

FOLD IN WICK TO REMOVE

BLACK LINE

UNFOLD WICK

ARROW MARK ON WICK MUST MATCH ARROW MARK ON STAINLESS STEEL WICK HOLDER

Fig. 5-14. Remove the old convection heater wick (courtesy Fanco USA).

Fig. 5-15. Install the new convection heater wick (courtesy Fanco USA).

adjuster and lift out the entire assembly (Fig. 5-13).

To remove the wick from the adjuster, fold it and slide it out (Fig. 5-14).

To install a new wick in a convection kerosene heater, see Fig. 5-15. After turning the wick adjuster counterclockwise as far as it will go, fold the new wick and slide it into the adjuster (Fig. 5-16). The line on the outside of the wick should match the bottom edge of the adjuster. In addition, the arrow mark on the wick

Fig. 5-16. Installing the wick (courtesy Kero Sun, Inc.).

must match the arrow mark stamped on the stainless steel wick holder.

Don't disassemble the stainless steel wick holder. It is very difficult to reassemble.

Check that the holes in the wick completely clear the rivets in the wick holder for smooth operation. Turn the wick adjuster clockwise as far as it will go. Check the height of the wick (Fig. 5-12), then press it against the teeth inside the adjuster to obtain a firm grip. Start at the bottom row of teeth and slowly work up in a circular motion. Check to see that the gasket is still in place on the burner tank. Replace the wick and adjuster mechanism on the burner tank. Make sure the wick fits evenly in place (Fig. 5-17).

Position the adjuster shaft to the front of the heater. Slowly work the assembly down. Do not catch the threads in the unwoven section of the wick on the wick holder mounting studs. After the assembly is in place, rotate 90 degrees rapidly left and right as well as up and down until the wick slides freely (Fig. 5-18). Recheck the wick height. Do not tighten wing nuts on the mounting studs. Rotate the wick adjuster counterclockwise as far as it will go. Using a 7/32-inch diameter drill bit as a gauge, check for correct clearance between the wick holder and the draft tube all around. If the clearance is not correct, loosen the wing nuts and adjust.

Fig. 5-17. Fitting the wick (courtesy Kero-Sun, Inc.).

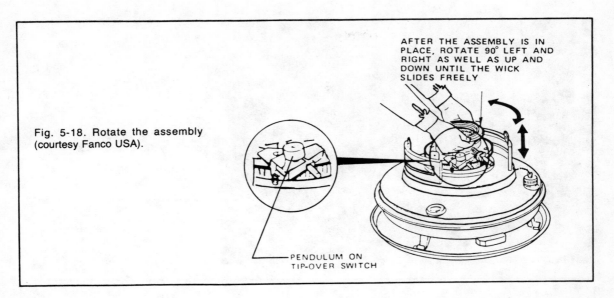

AFTER THE ASSEMBLY IS IN PLACE, ROTATE 90° LEFT AND RIGHT AS WELL AS UP AND DOWN UNTIL THE WICK SLIDES FREELY

PENDULUM ON TIP-OVER SWITCH

Fig. 5-18. Rotate the assembly (courtesy Fanco USA).

Tighten the wing nuts evenly in several steps. Recheck the draft tube and wick height measurements. Turn the knob clockwise and counterclockwise a few times to make sure the mechanism is functioning smoothly. Any ragged edges that appear at the top of the wick should be trimmed. Before reassembly, check the auto shufoff. Reset the system by pressing down on the set lever. Turn the wick adjuster knob clockwise as far as possible. Then upset the pendulum to trip the system. The wick should drop quickly to the lowest possible position.

Check the igniter before reassembly. Instructions for this job will be found later in this chapter.

Reinstall batteries, shelter plate, body base, wick adjuster knob, reset knob (on some models), outer cover, burner, top plate, and carrying handle. See Fig. 5-19.

IGNITER REPLACEMENT

As you learned in Chapter 2, the igniter or ignition coil is an automatic lighting mechanism that produces a spark to ignite the kerosene vapor when needed. The igniter can burn out or become inoperative and need replacement.

Replacing your igniter is a very simple process. Refer to Figs. 5-20 and 5-21.

Fig. 5-19. The wick in place (courtesy Kero-Sun, Inc.).

Fig. 5-20. Replacing the igniter (courtesy Robeson Industries Corporation).

Ignition coil

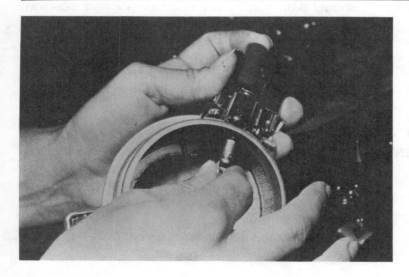

Fig. 5-21. Remove the ignition coil and check (courtesy Kero-Sun, Inc.).

116

Remove batteries from the case to insure that you won't be shocked or that the igniter won't accidentally ignite fuel while you're working on the unit. Never attempt to repair your portable kerosene heater while it is still warm.

Remove the guard and burner to gain access to the ignition mechanism. Push down the ignition knob. To remove the ignition coil, push in and turn counterclockwise. Install the new ignition coil and replace the burner, guard, and batteries. Before replacing the igniter, check the batteries to make sure that they are not low and thus the cause of poor ignition.

TROUBLESHOOTING YOUR HEATER

Though the concept and operation of the portable kerosene heater may be new to you, the following information can help you troubleshoot and repair your heater with minimal problems.

The biggest culprit in causing kerosene heater maintenance and repair problems is burning fuel with an incorrect—too high or too low—flame. Figure 5-22 illustrates what the flame should look like for a radiant heater, and Fig. 5-23 illustrates the same for a convection heater.

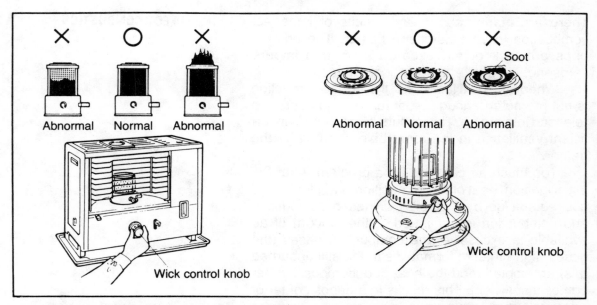

Fig. 5-22. Proper radiant heater flame adjustment (courtesy Fanco USA).

Fig. 5-23. Proper convection heater flame adjustment (courtesy Fanco USA).

Continuous operation with the wick lowered causes an extraordinary increase in the temperature of the wick and the upper face of the air supply tube. This, in turn, causes an increase in the temperature of kerosene in the fuel tank. Kerosene begins evaporating beyond the combustion capacity of the burner. The result is imperfect combustion. The fuel that has not been perfectly burned remains. This causes the flame to burst directly after extinguishing, and the flame leaks out of the hole in the wick casing.

As the raw fuel is liquefied again, kerosene appears at the wick-elevating shaft parts. The smell of kerosene results. As this problem continues, the top of the wick becomes stiff and causes additional problems: a damaged wick (especially jointed parts), malfunction of the wick drop mechanism, broken igniter coil, little burning, fire that extinguishes during use, burning noise (motorboating), and the smell of tar.

Correct adjustment of the flame through the wick control knob is the greatest preventer of problems with your kerosene heater. Correct adjustment can increase the safety and life of your heater while minimizing the cost of replacement and repair.

IMPERFECT COMBUSTION

There are many causes and results of imperfect combustion in your kerosene heater. The primary ones are lack of air, excessive air, and impure kerosene. Let's take a closer look at each.

When there is a lack of air, the oxidation reaction is not promoted enough, soot forms, and odors are released into the air. This occurs when there is insufficient ventilation in the room being heated by the heater.

Too much air can also be a problem. After the initial ignition, heat of decomposition is applied to the subsequent fuel by the fuel's own heat of combustion and combustion is continued. If the amount of air available is excessive, the combustion gases (the gases of the interim formations in the still unburned gas) are cooled, and the heat of decomposition becomes insufficient. This results in a deposit of tar or odors released to the air.

When kerosene with different ignition points,

such as white kerosene that has been left in direct sunlight for a prolonged period time, or white kerosene with quantities of other petroleum mixed in, the differences in the heat of decomposition will result in the clogging of the fiberglass wick, the formation of soot from imperfect combustion, and the release of odors.

QUICK TROUBLESHOOTER Based on what you've learned thus far about potential problems that can occur in portable kerosene heaters, let's develop the following quick troubleshooter.

Problem	Cause
Soot formation	Insufficient air
	Excessive fuel
	Mixed kerosene
Flames jumping	Excessive air
	Poor quality kerosene
	Water mixed with kerosene
Flame uneven	Insufficient air
	Excessive fuel
	Poor quality kerosene
Tar formation	Excessive air
	Insufficient fuel (wick too low)
	Mixed kerosene
Odors	Insufficient air
	Excessive air
	Insufficient fuel (wick too low)
	Excessive fuel
	Poor quality kerosene
	Mixed kerosene
	Water mixed with kerosene

Problem	Cause
No flame	Mixed kerosene Water mixed with kerosene
Flame goes out	Mixed kerosene Water mixed with kerosene

Here are quick cures for common troubles occurring in portable kerosene heaters.

Heater Won't Light

1. Water in kerosene.
   Remove and dry out the wick on absorbent paper. Drain the cartridge tank and burner tank. Wipe up spilled kerosene. Reinstall the wick. Fill with clean, pure K-1 low sulfur kerosene.
2. Empty tank.
   Fill the tank with clean pure K-1 low sulfur kerosene.
3. Igniter fails to glow properly.
   Check or replace batteries if weak or dead. Replace the igniter coil and disconnected or broken wires if batteries are good.
4. Igniter coil contacts side of wick.
   Use the wick adjusting knob to lower the wick until the igniter contacts the top of the wick.

Flame Flickers or Dies

1. Water in kerosene.
   Remove and dry out the wick on absorbent paper. Drain the cartridge tank and burner tank. Reinstall the wick. Fill with clean, pure K-1 low sulfur kerosene.
2. Wick covered with carbon or tar.
   Burn the wick clean. Trim the wick with scissors in more serious cases.

Flame Smokes or Causes Odor

1. Flame too high.
   Use the adjusting knob to lower the wick.

2. Air drafts hitting the heater.
   Move out of direct drafts.
3. Chimney not level.
   Use the wire handle to rotate the chimney until it seats on the ring encircling the wick.
4. Wick contaminated with carbon or tar.
   Burn the wick clean. Trim the wick with scissors in more serious cases.
5. Impure kerosene.
   Remove and dry out the wick on absorbent paper. Drain the cartridge tank and burner tank. Reinstall the wick. Fill with clean, pure K-1 low sulfur kerosene.

Excessive Wick Burning Down

1. Dangerous volatile fuel such as alcohol, gasoline, paint thinner, etc., mixed with kerosene. Drain the cartridge tank and burner tank. Replace the wick. Fill with clean, pure K-1 low sulfur kerosene.

Wick Adjuster Sticks

1. Water in kerosene.
   Remove and dry out the wick on absorbent paper. Drain the cartridge tank and burner tank. Reinstall the wick. Fill with clean, pure K-1 low sulfur kerosene.
2. Carbon or tar buildup on the wick.
   Burn the wick clean. Trim the wick with scissors in more serious cases.

Figures 5-24 through 5-27 offer a more detailed development of eight common symptoms of kerosene heater problems, their causes, and cures.

GENERAL INFORMATION   Maintaining your portable kerosene heater is simple with basic tools such as a couple screwdrivers and small wrenches. Kerosene heaters are relatively new to the American consumer, but they should not be intimidating. They are actually quite simple to understand and maintain. By following the logical troubleshooting guides and step-by-step repair instruc-

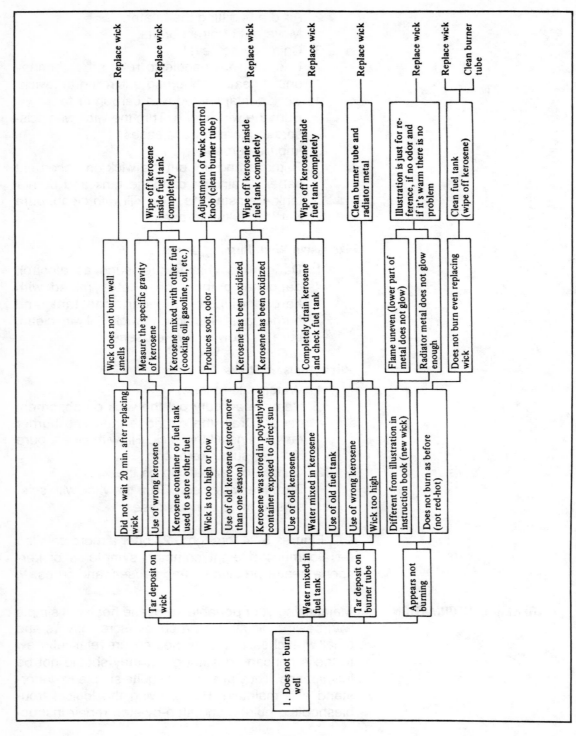

Fig. 5-24. Kerosene heater troubleshooting (courtesy Fanco USA).

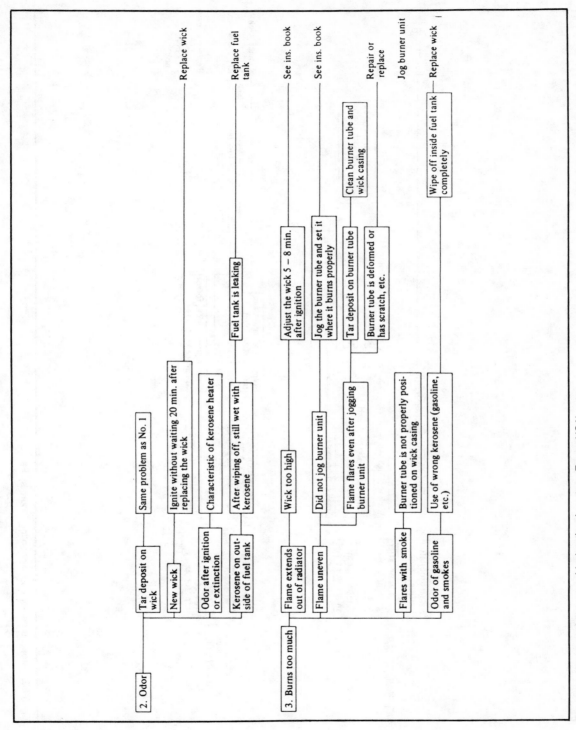

Fig. 5-25. Kerosene heater troubleshooting (courtesy Fanco USA).

123

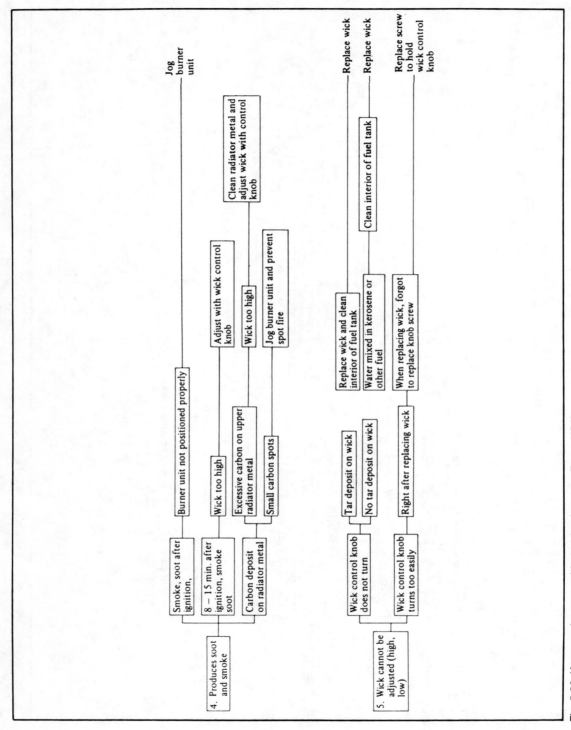

Fig. 5-26. Kerosene heater troubleshooting (courtesy Fanco USA).

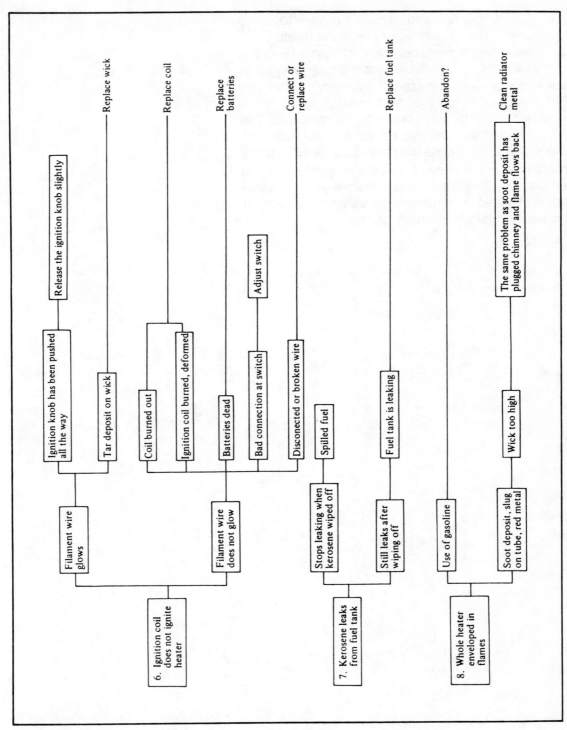

Fig. 5-27. Kerosene heater troubleshooting (courtesy Fanco USA).

tions in this chapter, you should be able to enjoy many warm winters with your kerosene heater.

Sometimes the most efficient thing to do is to call in an expert. Your kerosene heater dealer may be or may know of an authorized repair service that can offer parts and/or service as needed.

Check your heater's warranty to make sure that your maintenance activities do not void the manufacturer's warranty. Some warranties cover all repairs; others cover only a few. Read the warranty carefully and go over it with the dealer before you purchase your unit. Most important, think safety as you maintain your portable kerosene heater.

Chapter 6

Improving Kerosene Heater Safety

ARE PORTABLE KEROSENE HEATERS SAFE? UN-til recently, the answer has been "no." Kerosene heaters of the 1930s through 1950s were inefficient and smelly. They offered little or no safety features to reduce fire hazards.

After World War II in Japan, kerosene heaters were used extensively to heat homes. Competition for the marketplace bred innovative efficiency and safety features such as the secondary combustion burner. This unit burns the fumes from the first burner, which increases heating efficiency and reduces the kerosene odor from the air.

The chance of fire has been reduced by automatic mechanisms that retract or cover the wick if the unit is bumped or knocked over.

In the United States, kerosene heater safety has been greatly advanced by Underwriters Laboratories and their "Standard for Unvented Kerosene-Fired Room Heaters and Portable Heaters"—called UL 647. This publication outlines the requirements for receiving the UL listing on kerosene heaters that is *required* by many states on kerosene heaters sold there. The standard includes strict guidelines on general construction, electrical components, spacings, performance, and production tests. Figures 6-1 through 6-5 illustrate today's kerosene heaters.

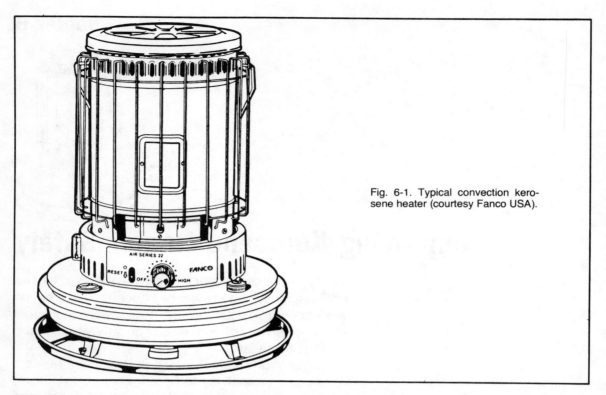

Fig. 6-1. Typical convection kerosene heater (courtesy Fanco USA).

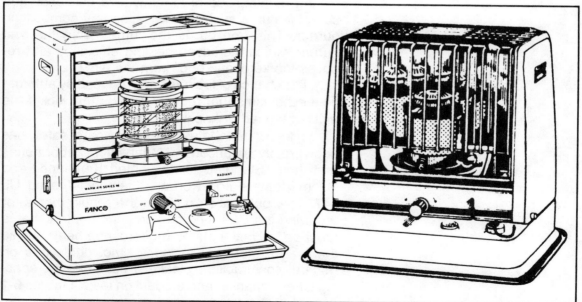

Fig. 6-2. Typical radiant kerosene heater (courtesy Fanco USA).

Fig. 6-3. Radiant heater with built-in tank (courtesy Kero-Sun, Inc.).

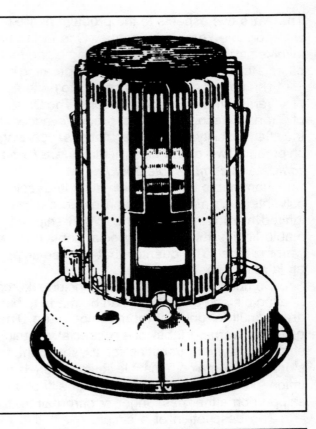

Fig. 6-4. Convection heater with built-in tank (courtesy Kero-Sun, Inc.).

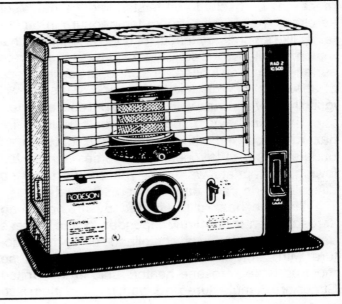

Fig. 6-5. Radiant kerosene heater with built-in fuel gauge (courtesy Robeson Industries Corporation).

There are two aspects to the problem of fire hazard with portable kerosene heaters. One is the hazard involved in the storage and handling of the fuel. The other is the hazard presented by an operating heater.

The nature of kerosene seems to be the subject of several common misconceptions. The Department of Transportation classifies products according to their flammability and sets regulations regarding their shipment. Two of the classifications are *flammable liquids* and *combustible liquids*.

Flammable liquids, which include gasoline and solvents such as acetone and ether, can easily be ignited at room temperature. When shipped, flammable liquids must be in special containers with a warning label. The quantity that may be shipped by air is restricted.

Combustible liquids, which include kerosene, heating oil, and diesel fuel, can be made to burn but are difficult to ignite at room temperature, They require no warning label or no special container for shipment. While quantities for air shipment are restricted for some combustible liquids, there is no limitation on the quantities of kerosene that may be shipped on either passenger or cargo aircraft.

The description of a simple qualitative experiment may be useful in understanding the nature of kerosene. Kerosene at room temperature in a container like an ordinary teacup cannot be ignited with a match. Similarly, a pool of kerosene on a concrete surface cannot be ignited. If a paper towel or a piece of absorbent cloth is used as a wick, kerosene will ignite readily and burn with a steady but docile flame.

The point being made is not that there is no fire hazard associated with kerosene, but that with an understanding of kerosene's properties and by following appropriate warnings, you should have no difficulty storing and handling kerosene safely.

The fire hazard presented by an operating kerosene heater is entirely dependent on the heater's design. You must rely on compliance of the manufacturer with safety standards such as UL 647. Because most portable kerosene heaters sold in the United States are manufactured in Japan, many of them also have approval from the Japan Heating Appliance Inspection Association. The Japanese standard is particularly stringent regarding the question of hazard under abnormal operation. There is a high incidence of earthquakes in Japan, and all kerosene heaters in that country must have an extinguishing mechanism that will snuff out the flame in the event of an earthquake or tip-over.

OXYGEN CONSUMPTION

Every flame-fired heating appliance requires oxygen. The amount of oxygen used is almost independent of the kind of fuel burned; it is proportional to the amount of heat produced. Thus, a typical wood stove rated at 50,000 Btu/hr. will use about five times as much oxygen as a kerosene heater rated at 10,000 Btu/hr. A typical residential furnace with a rating of 130,000 Btu/hr. will use 13 times as much oxygen as the kerosene heater.

Most houses have enough leakage around doors and windows and through cracks in the walls, floor, and ceiling to provide sufficient ventilation for furnaces. This leakage of fresh air in a house is called infiltration. It is recognized in standards for the installation of heating equipment. For example, the National Fire Protection Association (NFPA) standard No. 31 says, "In unconfined spaces in buildings of conventional frame, brick, or stone construction, infiltration normally is adequate to provide air for combustion and ventilation."

The standard defines a variety of furnace installation situations and recommends the size of ventilation openings to be provided when air is drawn directly from the outside and when it is drawn from adjacent rooms within the house.

This and other U.S. standards as yet don't make recommendations for ventilation of kerosene heaters, but the Canadian Standards Association (CSA) standard B140.9.3-M recommends that ventilation openings should be about four times larger per 1,000 Btu/hr. of rating for an unvented kerosene heater than for a vented furnace. This difference correctly reflects the fact that a vented heating appliance will have a somewhat greater tendency to draw air into a room than will an unvented one.

From the combined recommendations of these two standards, the oxygen requirements of a portable kerosene heater rated at 10,000 Btu/hr. will be amply provided for if a window to the outside is kept open by 1 inch, or if a door to adjacent rooms in the house is kept ajar by 1 inch.

To summarize, unvented kerosene heaters do use oxygen. If the oxygen is not replaced, these heaters could be hazardous. The label required by Underwriters Laboratories and information in the heater's owner's manual will indicate to you your heater's ventilation requirements.

CARBON MONOXIDE HAZARDS

Carbon monoxide is a poisonous gas and is common in our environment. Some sources such as fossil fuel power plants and automobiles are of concern because they contribute substantially to air pollution.

Other sources of carbon monoxide such as candle flames, kitchen gas stoves, and other combustion sources are not of concern because they do not significantly affect the environment. Deciding whether a particular source of carbon monoxide constitutes a hazard involves measuring the total amount of the gas.

A sensitive carbon monoxide tester will indicate significant concentrations of carbon monoxide around a candle flame. The total amount that the flame produces will not be enough to significantly affect the room's air quality. Similarly, the detection of carbon monoxide near a kerosene heater doesn't tell you whether or not this represents a hazard. A more thorough analysis of carbon monoxide hazards with unvented portable kerosene heaters is required. Let's look at the various standards that have been developed for testing this equipment.

There is no U.S. standard that addresses this problem for kerosene heaters. However, the American National Standards Institute has a standard for natural gas-fired cooking appliances for which test methods and performance requirements are similar. Donald S. Lavery, an analytical chemistry consultant, tested the Kero-Sun portable kerosene heater against these standards and found that carbon monoxide emission from these heaters is considerably lower than permitted by this standard.

Canada, Britain, and Japan have standards for kerosene heaters. The Japanese standard is considered the most stringent, and many models pass these standards before they are exported to the United States.

Flames tend to produce more carbon monoxide if the oxygen in their environment is depleted. This is of particular concern with unvented natural gas-fired burners and was an important reason why the Consumer Product Safety Commission (CPSC) proposed to ban unvented natural gas heaters. The proposal has since been withdrawn and replaced by a program to develop a mandatory standard. It is significant that the CPSC study of accident histories preceding the ban proposal led the commission to single out natural gas heaters from the proposed ban.

There are logical reasons why kerosene heaters are considered safer than gas-fired heaters. A wick-fed flame burns kerosene vapor as it evaporates from the wick. The rate of evaporation, and therefore the rate at which fuel is supplied to the flame, depends largely on the heat from the flame. Depletion of oxygen in the air supply causes a reduction in flame temperature, which reduces the fuel evaporation rate. The basic principle of a wick-fed burner is safer than that of a gas burner.

SAFETY RULES The safety of portable kerosene heaters can be improved by consumer awareness of the workings and warnings of these heaters. Now that you have a basic understanding of the hazards of kerosene heaters, let's consider important safeguards to operation and fuel handling.

The most important aspect of improving kerosene heater safety is your reading and understanding the owner's manual that came with your heater. In it you will find specific information on unpacking, setting up, lighting, and using your heater.

The following safety rules and guidelines should supplement rather than replace the safety information in your owner's manual.

Fuel

Never use anything but water-clear kerosene. Kerosene sold especially for heaters should have lower sulfur content. Refuel the heater outside and only after it has cooled 10 to 15 minutes. Store the fuel in clean, approved containers used only for kerosene.

Fire

Have a type B (B:C or A:B:C) fire extinguisher handy. Know how to use it. Be sure to have a working smoke detector. Every occupant should know what to do in case of fire. Don't use the heater near combustibles such as solids, liquids, and especially inflammable aerosols, vapors, or dusts.

Burns

The heater is hot. It will burn people, pets, or objects. Observe recommended clearances and keep children and pets away.

Ventilation

The following quote is from UL-647. "In a house of typical construction, that is, one which is not of unusually tight construction due to heavy insulation and tight seals against air infiltration, an adequate supply of air for combustion and ventilation is provided through infiltration. However, if the heater is used in a small room where less than 200 cubic feet of air space is provided for each 100 Btu per hour of heater reading (considering the maximum burner adjustment), the door(s) to adjacent room(s) should be kept open or the window to the outside should be opened at least 1 inch to guard against potential buildup of carbon monoxide. Do not use the heater in a bathroom or any other small room with the door closed."

Carbon Monoxide

Underwriters Laboratories points out that the allowed amount of carbon monoxide from a 10,000-Btu heater is less than that normally produced by a single burner on a gas kitchen range. Although little information is available on the effects of a heater's age and poor maintenance on carbon monoxide emissions, the ventilation recommendations have some margin of safety built in. The symptoms of carbon monoxide poisoning are flulike: nausea, headache, sore throat, and body pain. If you develop these symptoms, shut off the heater and ventilate the area.

Other Combustion Products

The effects of other heater emissions are being studied. Pollutants including carbon dioxide, nitrous oxides, sulfur oxides, formaldehyde, other gases, and fine particulates are emitted. Proper ventilation should avoid acute problems. Long-term chronic exposure to these substances at any level is unlikely to be beneficial.

The best safeguard is knowledge. Portable kerosene heaters should not be owned or operated by those unable or unwilling to follow the procedures in the owner's manual. Heaters should remain under the personal supervision of a competent operator while they are in use.

IMPORTANT SAFEGUARDS

Before you light your heater, make sure you have adequate ventilation. A door ajar from an adjoining room or a window opened slightly will be sufficient. Don't operate your heater in small, close quarters. Avoid any location where the heater would be in the path of a draft.

A kerosene heater should be kept a safe distance from any combustible materials such as clothing, draperies, furniture, etc. Three feet is the minimum spacing for a convection heater.

Keep children away from the heater when it is hot. Don't leave unsupervised children in a room where a heater is in use.

The heater will not function properly unless it is level. Make sure it is sitting on a level, hard surface rather than a thick carpet. Use a heat shield if possible.

Clean the outside of the heater regularly to remove dust or kerosene spilled when filling.

Make sure the heater is completely cool before touching it for any reason beyond adjusting the flame height or extinguishing the wick.

Again, use only clean, pure kerosene (K-1 or 1-K low sulfur). Old or discolored kerosene may cause an odor or smoke. Avoid contaminating fuel with water. *Never* substitute kerosene with gasoline (Fig. 6-6) or any other fuel. The portable kerosene heater is designed for *kerosene only*, and the use of anything else is extremely dangerous.

Don't try to fill the heater indoors in case any kerosene is spilled. Use a siphon to insure that fuel will not be spilled.

Store kerosene away from any heat (Figs. 6-7 and 6-8). Don't store it in polyethylene containers that can allow the fuel to deteriorate. Keep kerosene out of the reach of children.

Fig. 6-6. Never use gasoline in your kerosene heater (courtesy Fanco USA).

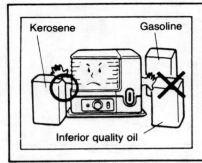  

Fig. 6-7. Use K-1 kerosene; never gasoline or inferior quality oil. Keep your heater and fuel can away from heat (courtesy Robeson Industries Corporation).

Don't attempt to carry or move the heater when it is lit. Decide where it should be placed before you light it. A kerosene heater should not be operated or stored in direct sunlight. The heat built up in the kerosene tank has a harmful effect on the fuel. The same goes for the kerosene storage can.

Place your kerosene heater only on a firm and stable platform (Fig. 6-9). The floor is the best location. Furniture or table finishes may be damaged by the heat generated. Place your heater out of the normal traffic pattern in your home—never in a hallway (Fig. 6-10). The heater may be accidentally

Fig. 6-8. Storing kerosene fuel (courtesy Robeson Industries Corporation).

GOOD

- Storage in a securely capped fuel can exclusively used for kerosene
- Storage in a cool as well as dark place that is free from any direct sunrays, rain, water, and fire

Fuel can

BAD

- Storage by means of plastic container
- Storage in a place (for example: porch or the like) that catches direct sunrays or rain

Plastic container

Fig. 6-9. Unsafe heater placement (courtesy Robeson Industries Corporation).

tipped over and, if the automatic extinguisher mechanism somehow jams, could spill kerosene onto the carpet or a rug that would serve as a perfect wick.

Don't attempt to repair a damaged heater yourself. Take or send it to an authorized repair station using the original packing material in the original carton. All fuel must be emptied beforehand.

Never remove the cartridge fuel tank while the unit is burning. Extinguish the flame of the kerosene heater and allow it to cool before the cartridge fuel tank is removed for any reason.

Never use your heater in areas where flammable vapor gases, dust, or clothing may be present (Fig. 6-11). Never store kerosene in any container other than an approved safety can.

Fig. 6-10. Never place your heater in a traffic area (courtesy Robeson Industries Corporation).

Fig. 6-11. Allow proper ventilation. Never hang combustibles near your heater (courtesy Robeson Industries Corporation).

ADDITIONAL SAFETY

Because of the pollutants and fire hazard, some people who would benefit from the high efficiency of the portable kerosene heater feel they cannot own one. Those with respiratory problems or with small children can install a radiant kerosene heater inside a fireplace, if they have one, and overcome these problems. An open fireplace flue can allow fresh air to replace burnt air while it gives pollutants an evacuation route. In addition, the fireplace screen can be used to keep little hands away from the hot elements of the kerosene heater while it is in operation.

If you plan to use your portable kerosene heater extensively throughout the heating season and particularly within one room, you might consider setting up a specific location for your heater. It must be out of the sunlight, away from traffic paths through the room and to adjacent rooms, in a location where ventilation is available, yet doesn't create a draft drawing heat from the room, and where the heat can be enjoyed by the most people. Then select your kerosene heater based on the size of the area you wish to heat. Keep an eye on your heater (Figs. 6-12 and 6-13).

THE ULTIMATE SAFETY MECHANISM

Safety in portable kerosene heaters is actually two types—design safety and operational safety. The kerosene heater is a relatively new device and is still

Fig. 6-12. Don't refuel a warm heater. Keep an eye on your heater to make sure it is burning efficiently (courtesy Robeson Industries Corporation).

Fig. 6-13. Check your kerosene heater to make sure it is off before leaving (courtesy Robeson Industries Corporation).

undergoing numerous changes and modifications for safety and efficiency as the competition for your heating dollar becomes keener. Manufactured safety is limited by the customer's desire to pay for it. We all want a theoretically 100-percent-safe heater, but few would be willing to buy it at the necessary $5,000 price. So we buy the model that offers the greatest safety features for under $300.

The difference in safety offered by the $5,000 and the $300 model can be taken up by responsible operational safety. The operator can follow the limitations and guidelines of the portable kerosene heater to make it as safe as possible.

This doesn't mean that additional progress in design safety should not be made. Instead, it means that until the "perfect" portable kerosene heater is offered for a "reasonable" price, you can safely and efficiently use your kerosene heater by learning its limitations and working within common sense safety guidelines. You are the ultimate safety mechanism.

FIRE PROTECTION

Fire hazards exist to some extent in nearly all houses. Even though the dwelling may be of the best fire-resistant construction, hazards can result from occupancy and the presence of combustible furnishings and other contents.

The following tabulation showing the main causes of fires in one- and two-family dwellings is based on an analysis of 500 fires by the National Fire Protection Association.

Cause of fire	Percent of total
Heating equipment	23.8
Smoking materials	17.7
Electrical	13.8
Children and matches	9.7
Mishandling of flammable liquids	9.2
Cooking equipment	4.9
Natural gas leaks	4.4
Clothing ignition	4.2

Cause of fire	Percent of total
Combustibles near heater	3.6
Other miscellaneous	8.7
Total	100.0

Fire protection engineers generally recognize that a majority of fires begin in the contents rather than in the building itself. Proper housekeeping and care with smoking, matches, and heating devices can reduce the possibility of fires. Other precautions to reduce the hazards of fires in dwellings—fire-stops, spacing around heating units and fireplaces, and protection over furnaces—are also recommended.

FIRE-STOPS Fire-stops are intended to prevent drafts that foster movement of hot combustible gases from one area of the building to another during a fire. Exterior walls of wood-frame construction should be fire-stopped at each floor level (Figs. 6-14 through 6-16), at the top-story ceiling level, and at the foot of the rafters.

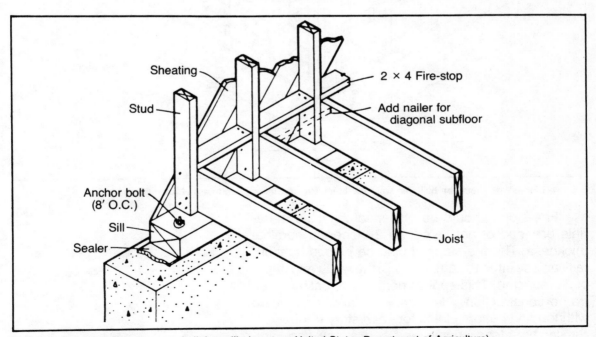

Fig. 6-14. Make sure fire-stops are built into sills (courtesy United States Department of Agriculture).

141

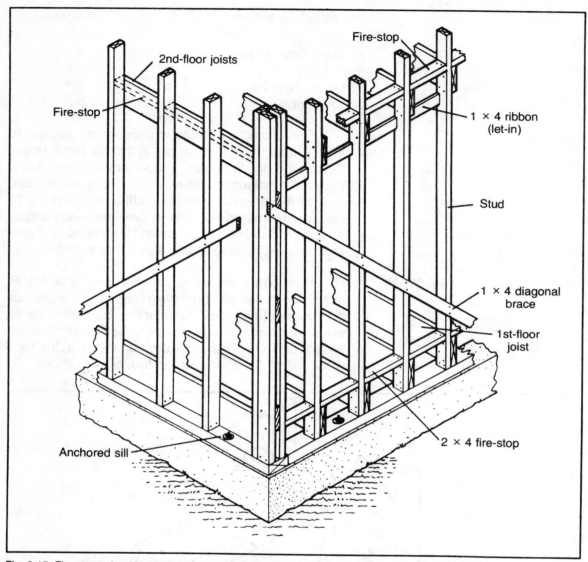

Fig. 6-15. Fire-stops should be built into walls (courtesy United States Department of Agriculture).

Fire-stops should be of noncombustible materials or wood of not less than 2 inches in nominal thickness. The fire-stops should be placed horizontally and be fitted to completely fill the width and depth of the spacing. This applies primarily to balloon type frame construction. Platform walls are constructed with top and bottom plates for each story (Fig. 6-17). Similar fire-stops should be used at the floor and ceiling of interior stud partitions, and headers should be used at the top and bottom of stair carriages (Fig. 6-18).

Noncombustible fillings should also be placed in any spacings around vertical ducts and pipes passing through floors and ceilings. Self-closing doors should be used on shafts such as clothes chutes.

When cold-air return ducts are installed between studs or floor joists, used portions should be cut off from all unused portions by tight-fitting stops of sheet metal or wood not less than 2 inches in nominal thickness. These ducts should be constructed of sheet metal or other materials no more flammable than 1-inch (nominal) boards.

Fire-stops should also be placed vertically and horizontally behind any wainscoting or paneling

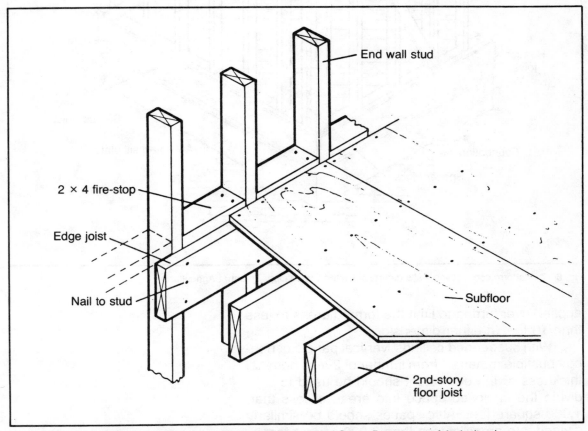

Fig. 6-16. Fire-stops should be built between floors (courtesy United States Department of Agriculture).

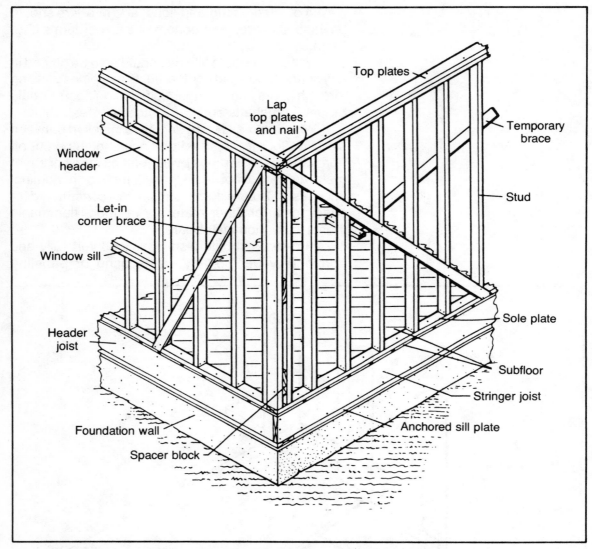

Fig. 6-17. Platform construction walls (courtesy United States Department of Agriculture).

applied over furring to limit the formed areas to less than 10 feet in either dimension.

With suspended ceilings, vertical panels of non-combustible materials from lumber of 2-inch nominal thickness or the equivalent should be used to subdivide the enclosed space into areas of less than 1,000 square feet. Attic spaces should be similarly divided into areas of less than 3,000 square feet.

144

CHIMNEY AND FIREPLACE CONSTRUCTION

The fire hazards within home construction can be reduced by insuring that chimney and fireplace constructions are placed in proper foundations and are properly framed and enclosed. Combustibles must not be placed too close to the areas of high temperature. Combustible framing should be no closer than 2 inches to chimney construction; however, the distance can be reduced to ½ inch when required, provided the wood is faced with a ½-inch-thick asbestos sheeting.

For fireplace construction, wood should not be placed closer than 4 inches from the back wall, nor within 8 inches of either side or top of the fireplace opening. When used, wood mantels should be located at least 12 inches from the fireplace opening.

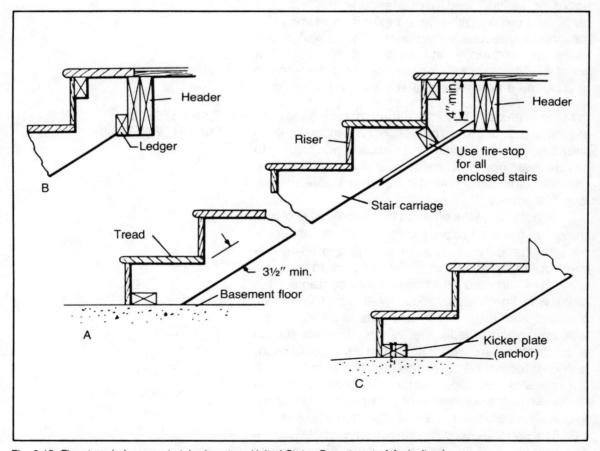

Fig. 6-18. Fire-stops in basement stairs (courtesy United States Department of Agriculture).

Faulty construction or improper use of heating equipment account for almost 25 percent of all fires. Many of these fires originate in the basement. Combustible products generally should not be located nearer than 24 inches from a hot air, hot water, or steam heating furnace. This distance can be reduced in the case of properly insulated furnaces or when the combustible materials are protected by gypsum board, plaster, or other materials with low flame spread. Most fire protection agencies limit to 170° F. the temperature to which combustible wood products should be exposed for long periods. Experimentally, though, ignition does not occur until much higher temperatures have been reached.

Added protection can be obtained with gypsum board, asbestos board, or plaster construction on the basement ceiling, either as an exterior surface or as backings for decorative materials. These ceiling surfaces are frequently omitted to reduce costs, but wood members directly above and near the furnace or a portable kerosene heater must be protected.

In some areas of a building, flame-spread ratings are assigned to limit the spread of fire on wall and ceiling surfaces. These requirements usually don't apply to private dwellings because of their highly combustible content, particularly in furnishings and drapes usually found in homes.

To determine the effect of the flammability of wall linings on the fire hazards within small rooms, burnout tests have been made. In one test, an 8-by-12-by-8-foot high room was furnished with an average amount of furniture and combustible contents. This room was lined with various wall panel products, plywood, fiber insulation board, plaster on fiberboard lath, and gypsum wallboard. When a fire was started in the center of this room similar to what could occur with a defective kerosene heater, the time to reach the critical temperature (when temperature rise became very rapid) or the flashover temperature (when everything combustible burst into flames) was not significantly influenced by either combustible or noncombustible wall linings. In the time necessary to

HEATING SYSTEMS

FLAME SPREAD
AND INTERIOR FINISH

reach these critical temperatures (usually less than 10 minutes), the room would already be unsafe for human occupancy.

Similar tests in a long partially vented corridor showed that the flashover condition would develop for 60 to 70 feet along a corridor ceiling within 5 to 7 minutes from the burning of a small amount of combustible contents. This flashover condition developed in approximately the same time, whether combustible or noncombustible wall linings were used, and before any appreciable flame spread along wall surfaces.

Wood panelling, treated with fire-retardant chemicals or fire-resistant coatings as listed by the Underwriters Laboratories, can also be used in areas where flame-spread resistance is especially critical. Such treatments, however, are not considered necessary in dwellings, nor can the extra cost of treatment be justified.

FIRE-RESISTANT WALLS

Whenever it is desirable to construct fire-resistant walls and partitions in attached garages and heating rooms, information on fire-resistance ratings using wood and other materials is readily available through local code authorities. Wood construction assemblies can provide ½-hour to 2-hour fire resistance under recognized testing methods, depending on the covering material.

Chapter 7

Installing Kerosene Heaters Permanently

KEROSENE HEATERS HAVE BEEN PORTABLE UNtil recently. They have been small and light enough to be moved from room to room within a home to heat wherever needed. They have been high-efficiency supplementary heaters. Primary residential heat came from more permanent systems such as gas, oil, coal, steam, and electrical heaters.

Recent innovations in the marketplace have brought kerosene heaters out of the portable world and into the permanent one. Why install a permanent kerosene heater? Kerosene heater technology has developed a highly efficient secondary burning system that both eliminates fuel odor and pulls nearly all available energy from the burning process. The permanent kerosene heater can work at 86 percent efficiency.

The permanent kerosene heater does not have the drawbacks of the portable heater. It cannot be set near combustibles by accident, it cannot be tipped over, it draws its air supply from the outside, and it can even draw fuel from safer tanks located at the exterior of the home. Some models offer computerized control that allows more efficient regulation of operating time and temperature.

The efficiency of your heating system affects your heating bill. The average seasonal efficiency of exist-

KEROSENE VERSUS OTHER FUELS

ing oil or gas central heating systems is 60 to 65 percent. Almost half of the heat they generate goes up the chimney or is dissipated through ducts and pipes on the way to heating your space. This is called *stack and transmission loss* and can be expensive.

Gas and oil central heat use inside air for combustion (Figs. 7-1 and 7-2). That air has to be re-

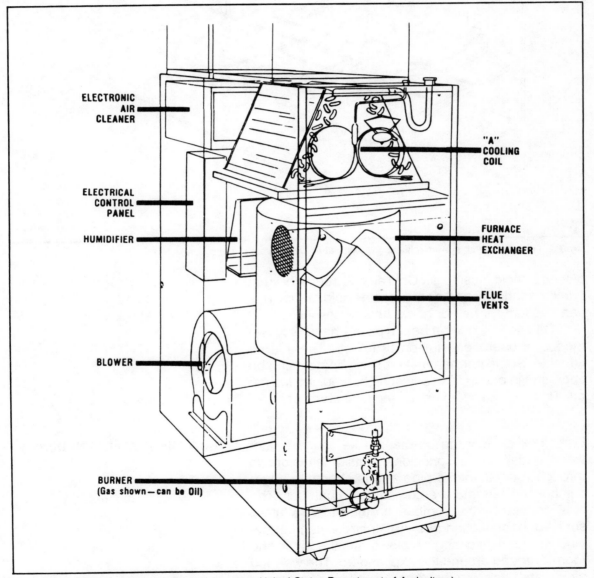

ELECTRONIC
AIR
CLEANER

ELECTRICAL
CONTROL
PANEL

HUMIDIFIER

BLOWER

BURNER
(Gas shown—can be Oil)

"A"
COOLING
COIL

FURNACE
HEAT
EXCHANGER

FLUE
VENTS

Fig. 7-1. Modern forced warm air furnace (courtesy United States Department of Agriculture).

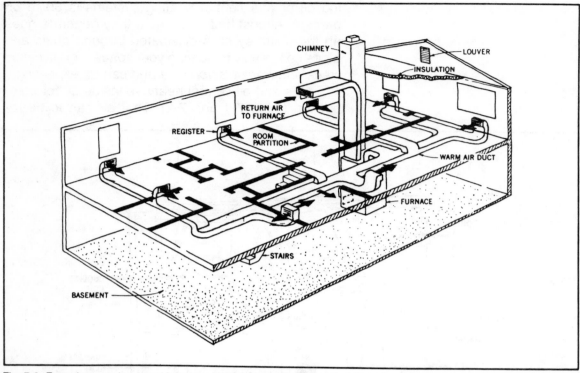

Fig. 7-2. Forced warm air system (courtesy United States Department of Agriculture).

placed by cold outside air. Cold air that seeps through cracks, windows, and doors creates cold drafts. This can account for half of home heating losses.

Gas and oil central heat typically have only two modes of operation—full-on or full-off. There is no variable. Some permanent kerosene heaters can be operating in one of four levels: high, medium, low, or off. This regulates fuel better and conserves energy.

HOW THE HEATER FUNCTIONS

Let's look at how the permanent kerosene heater works under normal conditions. Outside air is drawn through a small intake pipe and is channeled by a combustion fan into the combustion chamber. The heat created by the combustion then travels into a series of heat exchangers. An air circulation fan behind the heat exchangers blows air already in the room over and around the exchangers. The warmed air is sent back into the room.

APPLICATIONS

One of the original permanent kerosene heaters, the *Kero-Sun Monitor 20*, has a rating of 19,600 Btu/hr. on the high setting. It can be used as a secondary heat source in cooler areas and as a primary source in warmer areas. The unit consumes .156 gallons of fuel per hour at the high setting and, using the automatic controls, can run three winter days on 5 gallons of kerosene or six months on a 275-gallon tank.

Kero-Sun's newer and larger model, the *Monitor 30*, has a rating of 32,600 Btu/hr. on the high setting and consumes .24 gallon per hour. Five gallons will last two days. A 275-gallon tank will stretch refueling to 110 winter days. This unit can be a primary heat source throughout most of the United States. It will heat the typical 120-square-foot home in all but the coldest parts of the northern states.

Permanent kerosene heaters can be installed to back up the primary system and heat areas of the home such as upper rooms and basements. They can be installed efficiently in new additions to the home that weren't planned when the primary heating system was installed.

FUELING

There are three ways to fuel a permanent kerosene heater: by portable tank, gravity-fed tank, and pump. Portable tanks installed within the unit or, preferably, just outside the home are most practical for applications where little heat is needed or where the resident expects to move frequently. A 5- or 10-gallon tank is often installed to fuel a kerosene heater under these conditions.

Another choice is a large tank, typically a 55-gallon or 275-gallon one, available through kerosene heater and other heating fuel dealers. The advantage, besides the obvious one of less frequent refueling, is automatic home delivery.

If a gravity-fed tank is not feasible, you can install a tank below the level of your permanent kerosene heater, such as in the basement. A fuel pumping system can be used to bring kerosene up to the heater as needed.

Fueling of a permanent kerosene heater can often be easier than fueling a portable unit. Fuel companies and suppliers offer similar parts and services for users of fuel oil heaters.

INSTALLING A PERMANENT VENTED HEATER

Permanent vented kerosene heaters can be easily installed by even the novice do-it-yourselfer. The only requirements are a hole for the vent pipe and a plug for the electric fan and controls. Installation should only take an hour or two and can be accomplished with a 2½-inch hole saw or drill.

Unpack your permanent kerosene heater near the location where you expect to install it. You will probably need help bringing it into your home as it weighs about 100 pounds. Consider the best placement of your heater as the vent hole cannot be slid over to a new location. Check the outside of your home and make sure that the vent outlet will not be obstructed by objects.

Place the unit near the planned location and begin installation. Using a drill or saw, cut a 2½-inch hole (if appropriate) in the wall to the outside of the home. Install the vent pipe and attach it to the heater as shown in the instructions.

Drill a smaller hole for the fuel line and set the heater in place. Attach the line to the fuel source and to the heater unit.

Review the instructions that came with the heater to make sure that everything is right before testing it. When it is operating, check for leaks and problems before leaving it unattended for any length of time.

Permanent kerosene heaters can replace older heaters installed in the main part or basement of your home. Basement installation will probably require ducting to the main living area, which will reduce your heater's efficiency.

COMPARING FUEL COSTS

The fuels commonly used for home heating are wood, coal, oil, and gas. Electricity, though not a fuel, is being used increasingly. Kerosene, though used extensively during the 1930s, has not been a primary fuel source for many years because of its odor. Only with the introduction of secondary burners has kerosene begun to compete again with other fuels as a primary heating source for homes.

There are many considerations when comparing the cost of heating a home. The therms of heat per dollar should not be the sole consideration in selecting the heating fuel. Installation cost, the efficiency with which each unit converts into useful heat, and the insulation level of the house should also be considered. Electrically heated houses usually have twice the insulation thickness, particularly in the ceiling and floor, and may require considerably less heat input than houses heated with fuel-burning systems. To compare costs for various fuels, efficiency of combustion and heat value of the fuel must be known.

Heating units vary in efficiency, depending on the type, method of operation, condition, and location. Stoker-fired (coal) steam and hot water boilers of current design, operated under favorable conditions, have 60 to 75 percent efficiency. Gas- and oil-fired boilers have 70 to 80 percent efficiency. Forced warm-air furnaces, gas fired or oil fired with atomizing burners, generally provide about 80 percent efficiency. Oil-fired furnaces with pot burners usually develop not more than 70 percent efficiency. Permanent kerosene heaters as discussed in this chapter offer 86 to 92 percent efficiency.

Fuel costs vary widely in different sections of the country. For estimates, the data given in Table 7-1 can be used to figure the comparative costs of various fuels based on local prices. Here the efficiencies of electricity, kerosene, gas, oil, and coal are taken as 100, 85, 75, 75 and 65 percent, respectively. The efficiencies may be higher (except for electricity) or

Table 7-1. Information for Figuring the Comparative Costs of Various Fuels Based on Local Prices.

Fuel/Energy	One Therm Equals	Multiply by Cost	Equals
Electricity	29.3 kWH	per kWH	_____
Kerosene	.80 gallon	per gallon	_____
Gas, natural	127 feet	per feet	_____
Gas, LP	1.45 gallons	per gallon	_____
Fuel oil No. 2	.96 gallon	per gallon	_____
Coal	11.8 lbs. = .006 ton	per ton	_____

lower, depending on conditions. The values used are considered reasonable. The heat values are taken as 3,413 Btu/kWH of electricity for resistance heating, 136,000 Btu/hr. per gallon of kerosene, 1,050 Btu/cubic foot of natural gas, 92,000 Btu/gallon of propane or liquefied petroleum gas (LPG), 139,000 Btu/gallon of No. 2 fuel oil, and 13,000 Btu/pound of coal. A therm is 100,000 Btu.

To make a more accurate decision on whether to install a permanent kerosene heater, let's consider the benefits and disadvantages of other common heating sources.

WOOD

The use of wood requires more labor and storage space than other fuels. Wood fires are easy to start, burn with little smoke, and leave little ash.

Most well-seasoned hardwoods have about half as much heat value per pound as good coal. A cord of hickory, oak, beech, sugar maple, or rock elm weighs about 2 tons and has about the same value as 1 ton of good coal.

COAL

Two kinds of coal are used for heating homes—*anthracite* (hard) and *bituminous* (soft). Bituminous is used more often.

Anthracite coal sizes are standardized; bituminous coal sizes are not. Heat value of the different sizes varies little, but certain sizes are better suited for burning in firepots of given sizes and depths.

Both anthracite and bituminous coal are used in stoker firing. Stokers may be installed at the front, side, or rear of a furnace or boiler. Furnaces and boilers with horizontal heating surfaces require frequent cleaning because fly ash (fine powdery ash) collects on these surfaces.

OIL

Oil is a popular heating fuel. It requires little space for storing and no handling, and it leaves no ash.

Two grades of fuel oil are commonly used for home heating. Number 1 is lighter and slightly more expensive than No. 2, but No. 2 fuel oil has higher heat value per gallon. The nameplate or guidebook that comes with the oil burner indicates what grade oil should be used. Generally, No. 1 is used in pot burners and No. 2 in gun and rotary burners. Fuel oil should not be used in a kerosene heater, and kerosene should not be burned in fuel oil heaters.

Oil burners are of two kinds: *vaporizing* and *atomizing*. Vaporizing burners premix the air and oil vapor. The pot burner shown in Fig. 7-3 is vaporizing and consists of a pot containing a pool of oil. An automatic or handset valve regulates the amount of oil in the pot. Heat from the flame vaporizes the oil. In some heaters a pilot flame or electric arc ignites the oil pot when heat is required. In others the oil is ignited manually and burns continuously at any set fuel rate between high and low fire until shut off. There are few moving parts. Operation is quiet. Some pot burners can be operated without electric power.

Atomizing burners are of two general types: *gun* (or pressure) and *rotary*. The gun burner (Fig. 7-4 is by far the more popular type for home heating. It has a

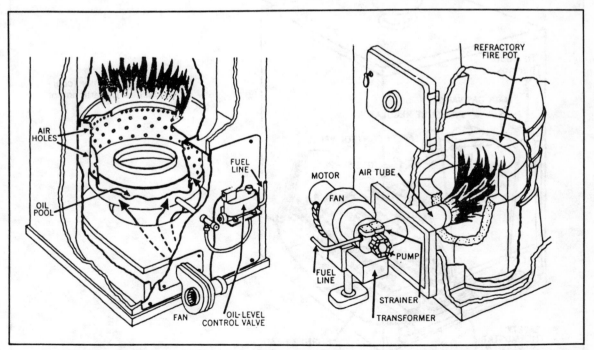

Fig. 7-3. Vaporizing or pot type oil burner (courtesy United States Department of Agriculture).

Fig. 7-4. Gun or pressure type oil burner (courtesy United States Department of Agriculture).

pump that forces the oil through a special atomizing nozzle. A fan blows air into the oil fog. An electric spark ignites the mixture, which burns in a refractory-lined firepot.

GAS

Gas is used in many urban homes and in some rural areas. It is supplied at low pressure to a burner head (Fig. 7-5), where it is mixed with the right amount of air for combustion.

A room thermostat controls the gas valve. A pilot light is required. It may be lighted at the beginning of

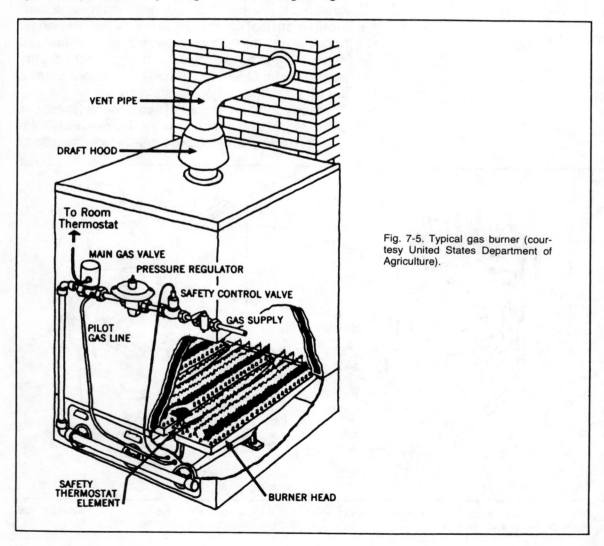

Fig. 7-5. Typical gas burner (courtesy United States Department of Agriculture).

the heating season and shut off when heat is no longer required. If it is kept burning during nonheating seasons, condensation and rapid corrosion of the system will be prevented. The pilot light should be equipped with a safety thermostat to keep the gas valve from opening if the pilot light goes out, so no gas can escape into the room.

Three kinds of gas—natural, manufactured, and bottled—are used. Bottled gas (usually propane) is sometimes called LPG for liquefied petroleum gas. It is becoming more popular as a heating fuel in recent years particularly in rural areas. Different gases have different heat values when burned. A burner adjusted for one gas must be readjusted when used with another gas.

Conversion gas burners may be used in boilers and furnaces designed for coal if they have adequate heating surfaces. Furnaces must be properly gastight. Conversion burners and all other gas burners should be installed by competent, experienced heating contractors who follow the manufacturer's instructions closely. Gas-burning equipment should bear the seal of approval of the American Gas Association.

Vent gas-burning equipment to the outdoors just as you would kerosene heaters. Keep chimneys and smoke pipes free from leaks. Connect all electrical controls for gas-burning equipment on a separate switch so that the circuit can be broken in case of trouble. Gas-burning equipment should be cleaned, inspected, and correctly adjusted each year.

Bottled gas is heavier than air. If it leaks into the basement, it will accumulate at the lowest point and create an explosion hazard. When bottled gas is used, make sure that the safety control valve is placed so that it shuts off the gas to the pilot and to the burner when the pilot goes out.

ELECTRICITY

Electric heating (Figs. 7-6 and 7-7) offers convenience, cleanliness, evenness of heat, safety, and freedom from odors and fumes. No chimney is required in building a new house, unless a fireplace is desired.

For electric heating to be more competitive

Fig. 7-6. Electric cable heat (courtesy United States Department of Agriculture).

Fig. 7-7. Electric wall heat (courtesy United States Department of Agriculture).

economically with other types of heating, houses should be well insulated and weather stripped, should have double or triple glazed windows, and should be vapor sealed. The required insulation, vapor barrier, and weatherproofing can be provided easily in new houses, but may be difficult to add to old houses.

The cost of electricity varies so greatly across the United States that only a localized estimate of heating costs between electric and kerosene heat can tell you which is most efficient.

Electric heating equipment should be only large enough to handle the heat load. Oversized equipment costs more and requires heavier wiring than does properly sized equipment.

INSTALLING A PERMANENT PORTABLE HEATER

There are many reasons for installing your portable kerosene heater permanently (Fig. 7-8). The first is safety. By locating your heater in a spot where it cannot be tipped over easily, you will minimize the possibility of fire. If you decide to install it in a vented area such as in a fireplace, you can also reduce the chance of accidental asphyxiation.

Before you decide to permanently install your

portable kerosene heater, make sure that you can refuel it easily. This may mean purchasing a heater with a removable fuel tank or having your heater modified with a fuel line originating outside the home.

If possible, mount your portable heater in a location that will completely eliminate spillage problems. Don't permanently install a heater on a carpet where spillage would allow the carpet to act as a wick. If possible, install the unit on a brick or firebrick base in a corner of the room where normal traffic will not disturb it. Because heat rises, you should install the unit as close to the floor as possible.

Because you are going to permanently mount your heater, you should install a fire extinguisher nearby. It should be a type B fire extinguisher or one with a B in the designation such as a B:C or A:B:C. The type B is designed to quickly extinguish petroleum fires.

Many people choose to install their portable kerosene heater permanently in an unused fireplace (Figs. 7-9 and 7-10). This installation offers numerous advantages over placing the heater in the room. First, greater ventilation is achieved through the chimney damper than comes from normal infiltration. Second, toxic fumes quickly escape to the outside. This is especially important if you have someone in

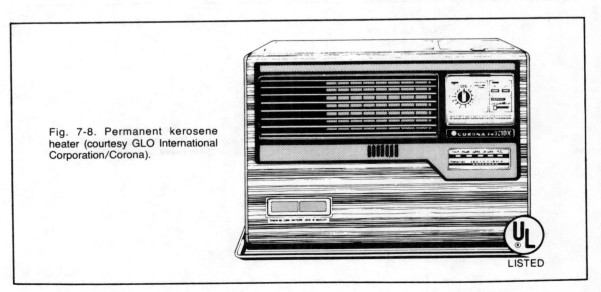

Fig. 7-8. Permanent kerosene heater (courtesy GLO International Corporation/Corona).

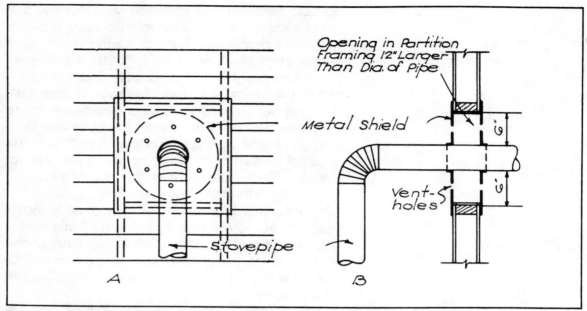

Fig. 7-9. Installing an externally-vented stovepipe (courtesy United States Department of Agriculture).

the home who suffers from respiratory problems that can be irritated by carbon monoxide or other combustion products. Third, the fireplace floor is designed to accept high heat without damaging surrounding elements.

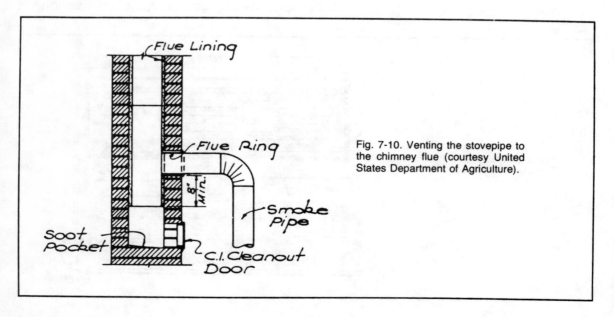

Fig. 7-10. Venting the stovepipe to the chimney flue (courtesy United States Department of Agriculture).

As noted earlier, only install a radiant kerosene heater in a fireplace. The reason is that a radiant heater directs the heat to the front of the unit while a convection heater radiates heat in all directions. Fireplaces are built to send heat toward the room. A convection heater would waste much heat that would be sent up the chimney. The radiant heater would reflect the heat out the firebox and into the room where the heat is needed.

There are numerous ways to install kerosene heaters permanently. Deciding to permanently install a kerosene heater can also depend on whether it will be a primary or secondary heat source and on what the other heat sources are. Most important, the kerosene heater must be installed safely to minimize both installation mishaps and future operating problems.

Chapter 8

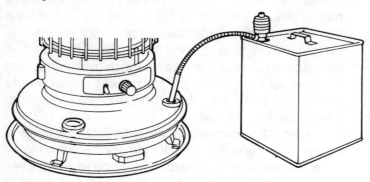

Selecting Kerosene Heater Accessories

NUMEROUS REPLACEMENT AND MODIFICATION PRO-
ducts are being offered to increase the benefits
of kerosene heating. Products available for the portable kerosene heater and how to make the best
selection from among them will be discussed (Figs.
8-1 through 8-3 and Tables 8-1 through 8-3). First,
let's consider the source.

As you shop around for your portable or permanent
kerosene heater, you will discover quickly that there
is a diverse selection of brands, models, features,
and claims. Fireplace shops, oil heat retailers, discount houses, alternative energy outlets, drugstores
and even supermarkets have kerosene heaters.

The merchants offering products for the secondary kerosene heater market are just as diverse.
Wicks can be found at drugstore checkout counters,
storage containers are available at auto parts stores,
and siphon pumps are stocked in clearance stores.
Where to buy?

Price is important. Kerosene heaters are purchased as economically efficient heating equipment
and, to stay that way, the replacement parts and
accessories must also be cost-efficient. Brand
names are important as they can offer consistent
quality. Companies invest millions of dollars in de-
veloping a brand name following. They cannot afford

PURCHASING ACCESSORIES

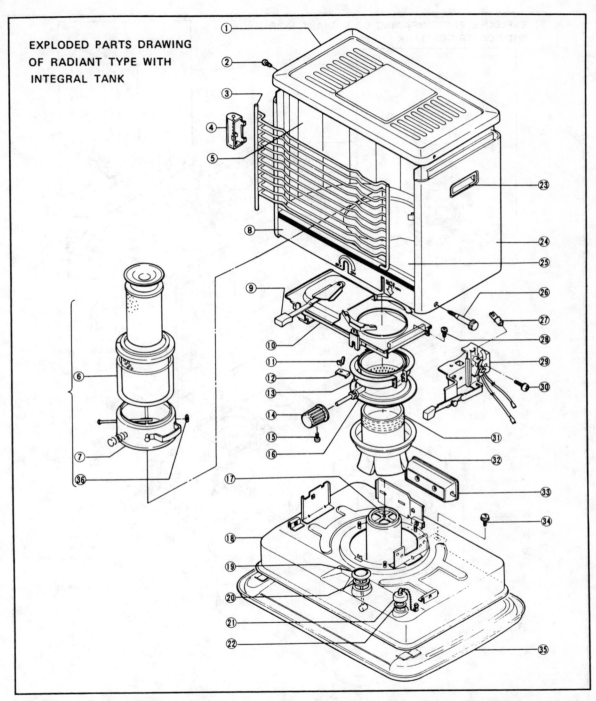

EXPLODED PARTS DRAWING
OF RADIANT TYPE WITH
INTEGRAL TANK

Fig. 8-1. Replacement parts are available from your dealer. Your owner's manual will have an exploded parts drawing (courtesy Fanco USA).

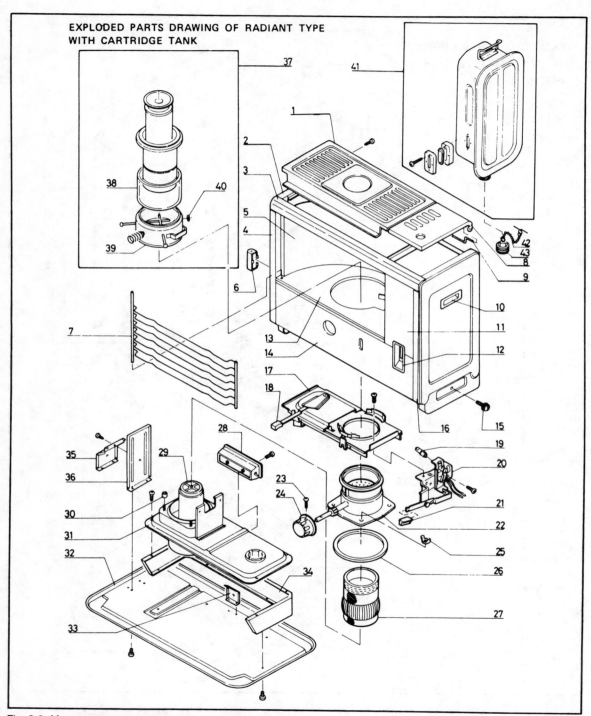

EXPLODED PARTS DRAWING OF RADIANT TYPE
WITH CARTRIDGE TANK

Fig. 8-2. Many parts are interchangeable between models, but verify this before replacement (courtesy Fanco USA).

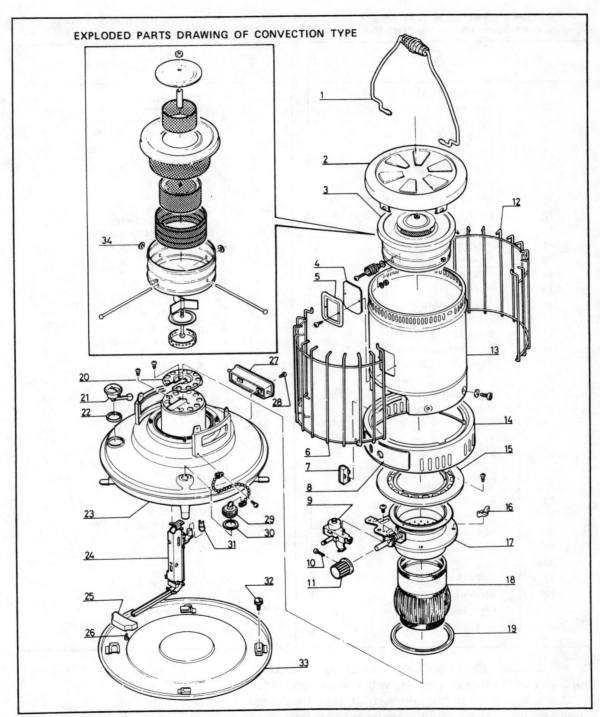

EXPLODED PARTS DRAWING OF CONVECTION TYPE

Fig. 8-3. Make sure you have an exploded parts drawing before disassembly (courtesy Fanco USA).

Table 8-1. Parts List for Radiant Heater with Integral Tank (courtesy Fanco USA).

PICTURE REFERENCE NUMBER	DESCRIPTION
1	TOP PLATE
2	TAPPING SCREW
3	GRILLE
4	LEVEL
5	REFLECTOR ASS'Y VERTICAL
6	GLASS COVER
7	CHIMNEY ASS'Y
8	PANEL
9	SAFETY RESET LEVER W/KNOB
10	AUTOMATIC SAFETY SHUTOFF
11	WING NUT
12	BRACKET - (K)
13	COVER BRACKET
14	KNOB
15	KNOB SCREW
16	WICK HOLDER ASS'Y
17	BURNER COVER ASS'Y
18	KEROSENE TANK ASS'Y
19	KEROSENE LEVEL GAUGE
20	RUBBER PACKING - (A)
21	FUEL CAP ASS'Y
22	RUBBER PACKING - (B)
23	HANDLE
24	BODY ASS'Y
25	REFLECTOR BOTTOM
26	THUMBSCREW
27	IGNITER COIL
28	SCREW
29	IGNITION ASS'Y
30	SCREW
31	WICK
32	RUBBER COVER
33	BATTERY CASE ASS'Y
34	KNURLED HEAD SCREW
35	DRIP TRAY
36	PUSH NUT

Table 8-2. Parts List for Radiant Heater with Cartridge Tank (courtesy Fanco USA).

PICTURE REFERENCE NUMBER	DESCRIPTION
1	TOP PLATE
2	BODY ASS'Y
3	ORNAMENT BAR
4	LEFT PANEL
5	REFLECTOR ASS'Y, VERTICAL
6	LEVEL
7	GRILLE
8	DOOR
9	SPRING HINGE
10	HANDLE
11	ORNAMENT PLATE
12	KEROSENE LEVEL WINDOW
13	REFLECTOR BOTTOM
14	FRONT PANEL
15	KNURLED HEAD SCREW
16	RIGHT PANEL
17	AUTOMATIC SAFETY SHUTOFF
18	SAFETY RESET LEVER W/KNOB
19	IGNITER COIL
20	IGNITION ASS'Y
21	IGNITION LEVER KNOB
22	WICK HOLDER ASS'Y
23	KNOB SCREW
24	WICK ADJUSTER KNOB
25	WING NUT
26	WICK HOLDER GASKET
27	WICK
28	BATTERY CASE ASS'Y
29	BURNER COVER ASS'Y
30	SPACER
31	COMBUSTION TANK ASS'Y
32	DRIP TRAY
33	REAR SIDE BRACKET
34	COMBUSTION TANK HOLDER
35	LEFT SIDE BRACKET
36	BRACKET FOR SHUTTER
37	CHIMNEY ASS'Y
38	GLASS COVER
39	GLASS HOLDER
40	PUSH NUT
41	CARTRIDGE TANK ASS'Y
42	CHAIN SET
43	FILL CAP ASS'Y
44	SEALANT FOR CARTRIDGE TANK PARTS

to sell shoddy merchandise. You should discuss purchases of kerosene heaters with owners and also review consumer publications that test heaters. These sources, though not completely accurate, can

PICTURE REFERENCE NUMBER	DESCRIPTION
1	HANDLE
2	TOP PLATE
3	BURNER UNIT
4	MICA
5	MICA FIXTURE
6	GUARD
7	LEVEL
8	PANEL
9	TIP-OVER SWITCH
10	KNOB SCREW
11	KNOB
12	GUARD
13	BODY
14	BODY BASE
15	SHELTER PLATE
16	WING NUT
17	WICK HOLDER ASS'Y
18	WICK
19	WICK CASING GASKET
20	VENTILATOR
21	KEROSENE LEVEL GAUGE
22	RUBBER PACKING - (A)
23	KEROSENE TANK ASS'Y
24	IGNITION ASS'Y
25	IGNITION KNOB
26	IGNITION KNOB SCREW
27	BATTERY CASE ASS'Y
28	TAPPING SCREW
29	FUEL CAP
30	RUBBER PACKING - (B)
31	IGNITER COIL
32	THUMB SCREW
33	DRIP TRAY ASS'Y
34	PUSH NUT

Table 8-3. Parts List for Convection Heater (courtesy Fanco USA).

help you make the most efficient purchase of both a kerosene heater and related products and accessories.

WICKS A good wick can make heater operation both safe and efficient. Conversely, a poorly designed or manufactured wick can cause untold problems with carbon buildup, poor secondary burning, and toxic fume release. It also increases the danger of fire.

Most kerosene heater wicks are of fiberglass tops and cotton or wool bottoms. Why two fibers? Fiberglass provides efficiency, clean burning, and long life. Cotton or wool bases assure absorption and an even flame. The fiberglass part of the wick is made of special glass fibers and carbon fibers with consideration given to strength and workability. This wick causes the kerosene to vaporize from the surface of the very thin glass fibers and sends the kerosene vapor to the combustion section.

The lower part of the wick of cotton or wool fibers has excellent moisture absorption properties. It lifts the kerosene in the tank to the fiberglass wick using the capillary tube effect.

Select the wick that is most appropriate to your type of portable kerosene heater. Some manufacturers will void the warranty of their heater if the incorrect type of wick is installed (Fig. 8-4). Check your heater's warranty and talk with the dealer who sold you your heater to find out what is recommended.

Remember that off-brand wicks sold at discount prices can be a very poor investment. They cause the release of toxic fumes and may avoid the automatic extinguishing mechanism's effort to squelch the fire in case of an accident (Fig. 8-5). Certain models or types of portable kerosene heaters will have slightly different wicks. The replacement wick should be identical to the original wick for safety and efficiency.

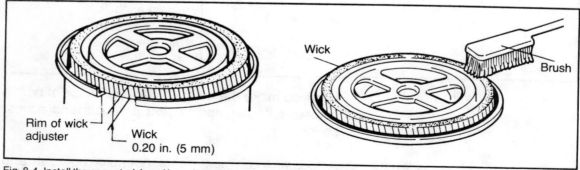

Rim of wick adjuster

Wick 0.20 in. (5 mm)

Wick

Brush

Fig. 8-4. Install the correct wick and keep it adjusted correctly (courtesy Robeson Industries Corporation).

Fig. 8-5. Keep the wick clean (courtesy Robeson Industries Corporation).

IGNITERS

Igniters are battery-operated ignition mechanisms that are designed to spark when the control button is pushed or turned. This spark ignites the kerosene fumes rising from the wick and creates the flame.

Igniters are simple units and are the second-most replaced unit on the portable kerosene heater. Chapter 2 offers step-by-step instructions on how to install or replace an igniter in both radiant and convection kerosene heaters.

Figure 8-6 illustrates the components within the typical igniter: filament, stem, protector, base, and glass. Figure 8-7 shows what normal and abnormal filaments will look like in an igniter.

As with wicks, quality is important. It is less critical, though, as the igniter or ignition coil is only used to start and not to continuously operate the kerosene heater. If required, the igniter can be safely overridden by a safety match.

SIPHON PUMPS

If your portable kerosene heater did not come with a siphon pump, get one. As you attempt to fuel your heater, you will find that pouring kerosene from a can into a funnel installed in the fuel tank is a difficult and dangerous task. It also gets boring if you have to do it each day during the winter season. It's better to spend a few extra dollars on a siphon pump that will insure safe and easy refueling of your heater.

There are three types of siphon pumps available for use in refueling kerosene heaters: manual, electric, and electric with auto shutoff. The manual siphon pump is the simplest and least expensive—usually

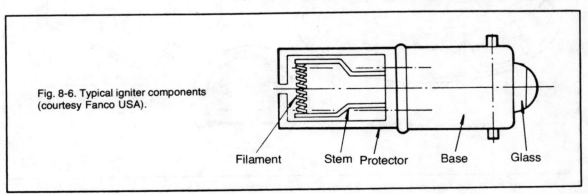

Fig. 8-6. Typical igniter components (courtesy Fanco USA).

Filament   Stem Protector   Base   Glass

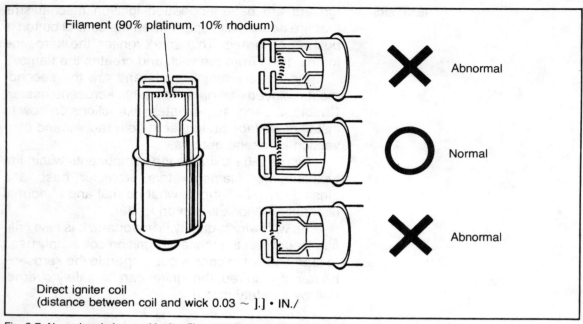

Filament (90% platinum, 10% rhodium)

Abnormal

Normal

Abnormal

Direct igniter coil
(distance between coil and wick 0.03 ~].] • IN./

Fig. 8-7. Normal and abnormal igniter filaments (courtesy Fanco USA).

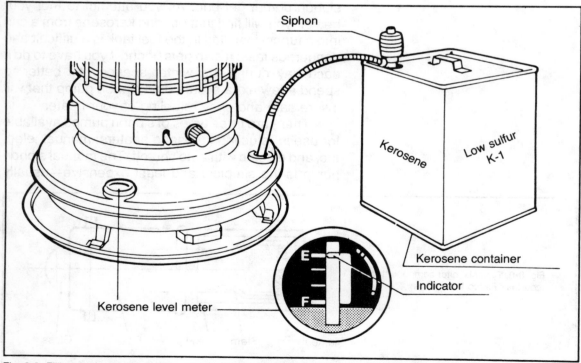

Siphon

Kerosene

Low sulfur K-1

Kerosene container

Indicator

E

F

Kerosene level meter

Fig. 8-8. Filling the built-in tank with a siphon (courtesy Robeson Industries Corporation).

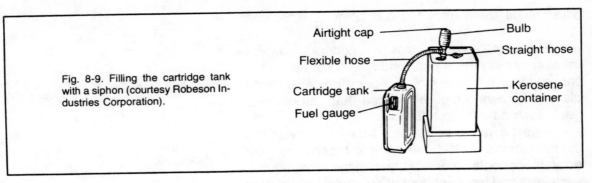

Fig. 8-9. Filling the cartridge tank with a siphon (courtesy Robeson Industries Corporation).

costing less than $5. It is operated by compressing a bulb or plastic chamber that draws the kerosene from the container. Releasing the chamber then forces the liquid into the tank (Figs. 8-8 and 8-9). The electric unit operates without the manual pumping. The third type is also electric and will automatically sense when the level of the fuel has reached the top of the container or tank. The pump will stop pumping to insure that it does not overflow. This pump costs $20 to $25.

STORAGE CONTAINERS

Many state and local governments limit the amount of gasoline or kerosene you can store in portable storage containers. Before you purchase such a container, check with local fire officials and government authorities to make sure that you're legal. Talk with your insurance agent to make sure that fuel storage does not modify your insurance policy. If you live in a rental home or apartment, check with the landlord or owner.

Storage containers come in various sizes from 1 quart to 55 gallons and more. The most popular with kerosene heater owners is the 5-gallon can. Some cans have special linings to resist corrosion. All cans, by law and by practicality, should be clearly marked KEROSENE. The danger of fueling a kerosene heater with gasoline has already been discussed. It can be highly explosive. Keep kerosene in KEROSENE cans and gas in GASOLINE cans. You might know which is in which without reading the label, but someone else might not.

The National Kerosene Heater Association suggests that kerosene containers be blue in color to differentiate them from the red color of most gas cans.

Store the kerosene cans in a cool place where the sun moving through a window will not fall on the can and warm it to a high temperature. Also, don't place a kerosene can in an area that can develop heat, such as next to a furnace.

Select a kerosene supplier that you know offers the highest quality K-1 low sulfur kerosene fuel with no water or contaminants. Buy from the same source each time and feel confident about the quality of your fuel.

Portable reflectors can be used to direct the heat from kerosene heaters to specific areas of a room to increase efficiency. When selecting a reflector, make sure that it fits your model heater and can do the job you need. A jerry-rigged reflector can cause a fire hazard.

Some kerosene heaters are designed to double as light or cooking heat sources (Fig. 8-10). Cooking rails are available for these models to allow you to warm a can of pork and beans or fry a trout. These

OTHER ACCESSORIES

Fig. 8-10. Kerosene light and heat source (courtesy Kero-Sun, Inc.).

cooking rails reduce the chance of spills as they increase the versatility of your portable kerosene heater.

Tool kits are standard with some kerosene heaters and optional with others. An adequate tool kit for these heaters includes a wick cleaning tool, metal brush, brass scraper tool, gauge for setting and checking wick height, a minisiphon for emptying the fuel tank, a nonabrasive cleaning pad, a cotton flannel polishing cloth, and related materials. It may also include standard workshop tools such as a screwdriver set, special-sized wrenches, and other tools to make maintenance and repair simpler.

Deodorizers are also available for use with some brands and models of portable kerosene heaters. Ceramic catalytic and other types of deodorizers are designed to reduce shutoff odors that occur in all kerosene heaters. These units can be purchased with or without installation hardware. Select one manufactured expressly for your brand and model to insure proper fit.

Additional accessories available now or soon are safety outriggers to make the heaters more stable, automatic extinguishers, heater platforms, and ducting. Man's ingenuity is often displayed in the accessories offered for primary products such as the portable kerosene heater.

HEATER FANS

Another accessory that you should consider in the purchase of a portable kerosene heater is a fan. It's actually an option available on many models. Some heaters with the boxy look of radiant kerosene heaters are actually convection heaters with reflectors and a fan to more efficiently direct the heat being produced. The fan draws about 50 watts of electric power in a typical model. The heater must be placed close to an electrical outlet, but this limitation is usually no problem.

OTHER APPLICATIONS

At least one manufacturer offers a portable kerosene cooking stove that weighs about 14 pounds. Its rated output is more than 7500 Btu and can burn 10 to 14 hours on a refueling. It is recommended for outdoor use such as on a boat or at a campsite. The stove can also be used indoors to warm fondues or help in canning.

Another model provides both heat and light. It uses a glass mantel surrounding the combustion chamber. The light is enough to read by, the heat is rated at over 8500 Btu/hr., and accessories are available for cooking on top of the unit.

Portable kerosene heaters are also being used in other applications where heat is important. Hobbyists are using these heaters to dry glues in shops and on models. Commercial painters are using them to quickly dry paints as they move from room to room. Mechanics are installing them in unheated garages where they work. Fishermen are taking them aboard boats. Recreation vehicle owners are using them in travel trailers and motor homes as cost-efficient replacements for traditional propane and LPG heaters.

Chapter 9

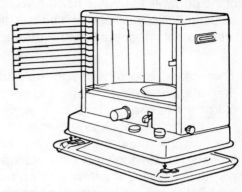

Buying a Used Kerosene Heater

THE PORTABLE KEROSENE HEATER HAS ONLY BEEN in America for a few years. The number of these heaters has grown tremendously. Many used or secondhand kerosene heaters will become more prominent in the marketplace.

USED HEATER SOURCES

You may decide that you can purchase a used kerosene heater for less than a new one. This may be so, but there are other elements to consider before cash changes hands. The first is the source.

The who and why of buying kerosene heaters or any other consumer product are important. A heater dealer who has reconditioned the unit and stands behind it with a limited warranty is a better source than the garage sale merchant in a distant city. The dealer is going to want more for his unit than the garage sale merchant. In either case, the best advice is to know what you are buying, from whom, and why they are selling.

A retailer or reconditioner of heaters depends on repeat sales or referrals to increase business. Satisfaction, within reason, must be guaranteed. The retailer must offer a useful and profitable product. The price must reflect both characteristics.

The garage sale merchant doesn't worry about repeat business. All merchandise is sold as is in most cases. There are no guarantees. The price reflects this in most cases. If you have a basic understanding of the portable kerosene heater, you can shop garage sales and flea markets for a heater. Inspect the heaters and knowledgeably bid on them.

KNOWING WHAT YOU BOUGHT

There is more to knowing portable kerosene heaters than understanding the steps to operation. You must study the technical side of these heaters as represented in this book and owner's manuals. Review the owner's manual that comes with the heater. If there isn't one, don't buy the heater as different models have individual safety and operating points that are vital. If necessary, you may be able to purchase a manual from a dealer or directly from the manufacturer's distributor. Make sure that is for the *same* model.

Figures 9-1 and 9-2 illustrate the exploded views of two different models of portable kerosene heaters taken from their operating manuals. This information is vital in inspecting and replacing damaged or missing parts. It is also useful in discovering what safety features are included in the unit. The owner's manuals will also give you instructions on ordering replacement parts.

INSPECTING THE HEATER

Let's go through a typical inspection of a used kerosene heater to decide whether it is worth purchasing and repairing. Not all owners will allow you to do this until you have bought the unit. You may want to get some kind of written agreement that allows you to return for full credit any unit that is defective.

The initial inspection includes looking at the exterior of the unit for obvious signs of damage or mishap. This could be a dent where the unit was dropped or a burn mark where heat has damaged the finish. Check for a broken wick adjuster knob, fuel level indicator, tank, or grille.

Remove the unit cover. Figure 9-3 illustrates how the cover is removed on a typical radiant heater. Carefully lift the cover in case loose parts fall out. Make this inspection at a workbench rather than in the home.

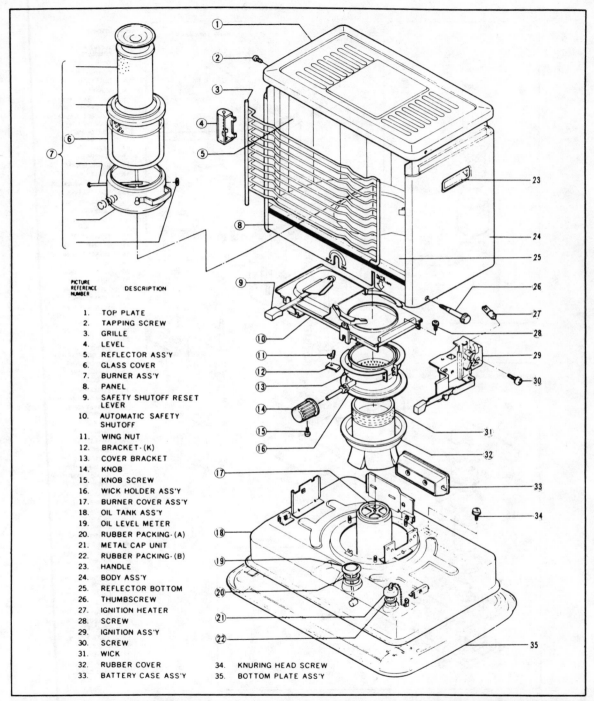

PICTURE
REFERENCE
NUMBER   DESCRIPTION

1.   TOP PLATE
2.   TAPPING SCREW
3.   GRILLE
4.   LEVEL
5.   REFLECTOR ASS'Y
6.   GLASS COVER
7.   BURNER ASS'Y
8.   PANEL
9.   SAFETY SHUTOFF RESET
      LEVER
10.   AUTOMATIC SAFETY
      SHUTOFF
11.   WING NUT
12.   BRACKET- (K)
13.   COVER BRACKET
14.   KNOB
15.   KNOB SCREW
16.   WICK HOLDER ASS'Y
17.   BURNER COVER ASS'Y
18.   OIL TANK ASS'Y
19.   OIL LEVEL METER
20.   RUBBER PACKING- (A)
21.   METAL CAP UNIT
22.   RUBBER PACKING- (B)
23.   HANDLE
24.   BODY ASS'Y
25.   REFLECTOR BOTTOM
26.   THUMBSCREW
27.   IGNITION HEATER
28.   SCREW
29.   IGNITION ASS'Y
30.   SCREW
31.   WICK
32.   RUBBER COVER
33.   BATTERY CASE ASS'Y

34.   KNURING HEAD SCREW
35.   BOTTOM PLATE ASS'Y

Fig. 9-1. Before buying a used kerosene heater, make sure the operating manual is included and contains a parts list (courtesy Robeson Industries Corporation).

177

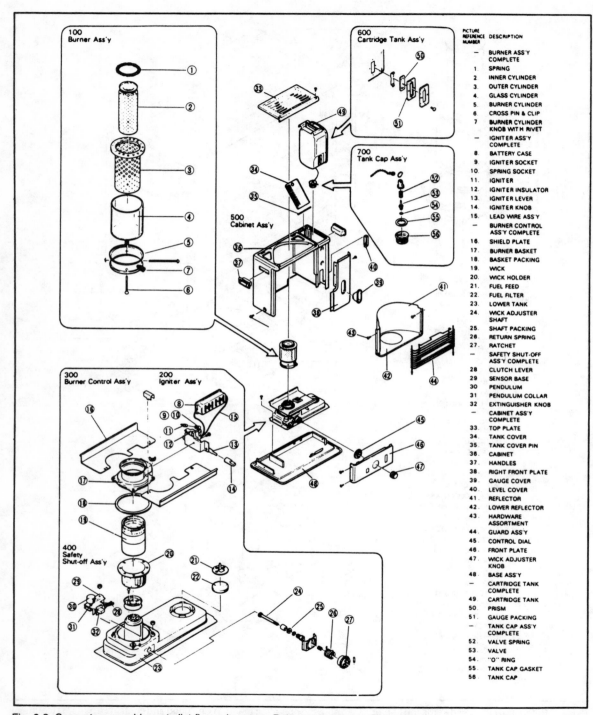

Fig. 9-2. Separate assembly parts list figure (courtesy Robeson Industries Corporation).

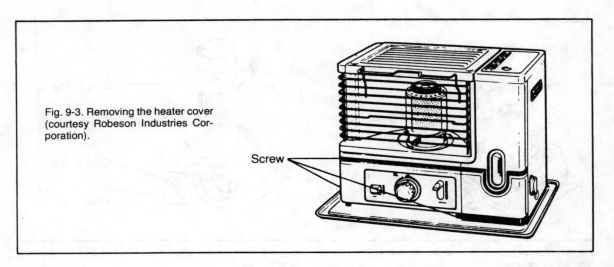

Fig. 9-3. Removing the heater cover (courtesy Robeson Industries Corporation).

Screw

Remove the tank, burner (Fig. 9-4) and any batteries installed in the heater. Remove the wing nuts fastening the burner basket (Fig. 9-5) for a closer inspection of the wick and burner assembly.

Pull the wick holder upward and remove the used wick (Fig. 9-6). Remember that the wick is not supposed to burn in a kerosene heater. It is the vapor emitted by the wick that burns. Inspect the wick for burns, buildup, or other damage that would indicate incorrect fuel or use.

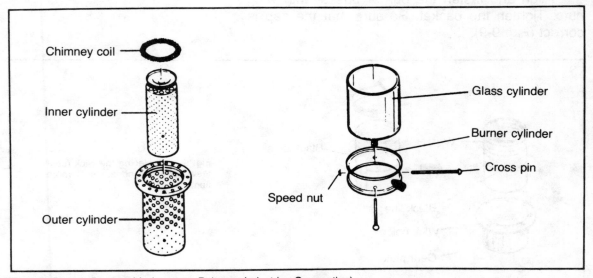

Chimney coil

Inner cylinder

Outer cylinder

Glass cylinder

Burner cylinder

Cross pin

Speed nut

Fig. 9-4. Burner disassembly (courtesy Robeson Industries Corporation).

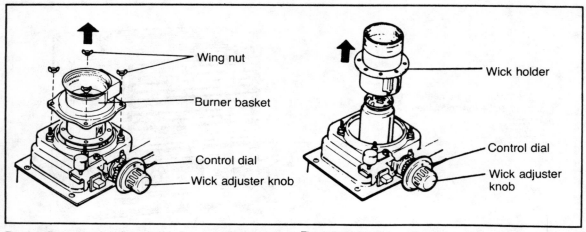

Fig. 9-5. Burner basket disassembly (courtesy Robeson Industries Corporation).

Fig. 9-6. Wick holder removal (courtesy Robeson Industries Corporation).

Figure 9-7 illustrates how to replace the old wick with a new one. For safety and efficiency, it's best to replace both the fuel and the wick on a used unit.

Proper operation of the extinguishing mechanism is also vital to the safety of the heater. Using instructions in the owner's manual, make sure that the automatic extinguishing mechanism works properly before using the unit. Figure 9-8 illustrates how the model unit works.

Align and fasten the burner basket with wing nuts. Tighten the basket. Be sure that the gap is correct (Fig. 9-9).

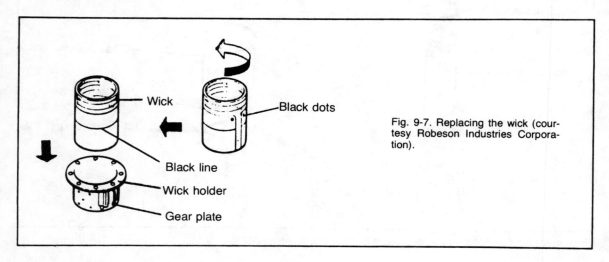

Fig. 9-7. Replacing the wick (courtesy Robeson Industries Corporation).

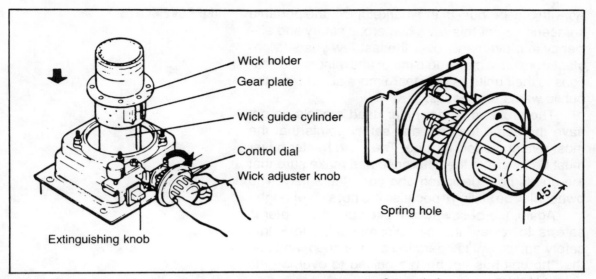

Fig. 9-8. Checking the extinguisher mechanism (courtesy Robeson Industries Corporation).

Before refueling, tip the unit to test the automatic extinguishing mechanism. Make sure that the flame would be extinguished by this action.

Finally, refuel and test the unit according to the instructions in the owner's manual. If you suspect that the incorrect fuel has been used in the unit, flush the tanks with pure K-1 low sulfur kerosene and refill. Wait about 20 minutes for the kerosene to work its way back up the wick before lighting. Check the ignition coil for spark and smooth operation.

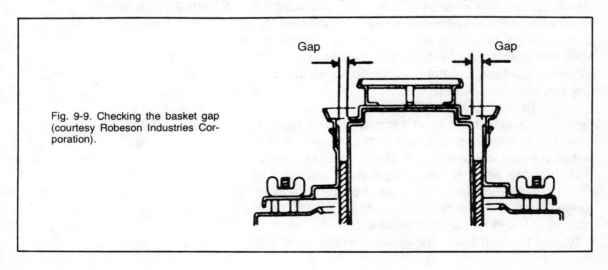

Fig. 9-9. Checking the basket gap (courtesy Robeson Industries Corporation).

As with any product in the marketplace, the portable kerosene heater has seen numerous safety and efficiency improvements over the last few years. Manufacturers have made both major and minor modifications to their units as they learn more about what the public wants and needs.

The problem is that older used models do not have the same features and safety points that the most recent models include. The used heater buyer must be aware of the difference and make sure that the unit is both designed and operated safely. The owner may be selling it because it is not safe enough.

Again, the best way to make sure the heater is safe is to review the owner's manual, check that safety equipment is operating as designed, and see that the unit has not been modified to override the safety features.

Be sure that the unit you purchase is legal. Many states and cities require that portable kerosene heaters sold and operated in their jurisdiction meet certain requirements, often listing by Underwriters Laboratories. Some codes may require that the unit be inspected. Your insurance company may have requirements so as not to void your homeowner's or renter's policy. Your landlord may have rules against the operation of portable kerosene heaters.

STORING YOUR HEATER

When the heating season is over, you will want to store your portable kerosene heater so it will be safe and ready for operation next winter. These instructions are specifically for a radiant heater with a removable fuel tank, but they can be easily adapted to any portable kerosene heater.

Agitate (Fig. 9-10) and wash the inner part of the tank with a little remaining kerosene, then pour it entirely out. Water rarely mixes with kerosene, and it will cause rust inside the tank. Remove all kerosene from the tank and dry the inside if possible.

Take the fuel tank out of the kerosene heater. Ignite and keep the wick burning. When the red heat of the outer cylinder becomes faint, turn up the wick fully and leave it there for about an hour until the

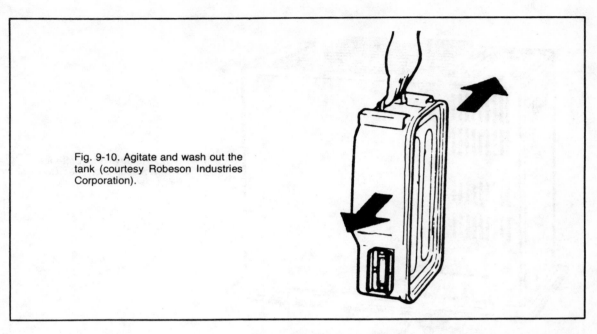

Fig. 9-10. Agitate and wash out the tank (courtesy Robeson Industries Corporation).

kerosene heater extinguishes itself. This will remove the fuel and the carbon from the wick.

Let the unit cool. Remove the burner and the batteries. Detach the cabinet. Remove the burner basket from the fuel reservoir and dry the inside of the reservoir thoroughly. Remove the carbon accumulated on the burner basket and/or soot adhering to the burner with a brush or screwdriver (Fig. 9-11).

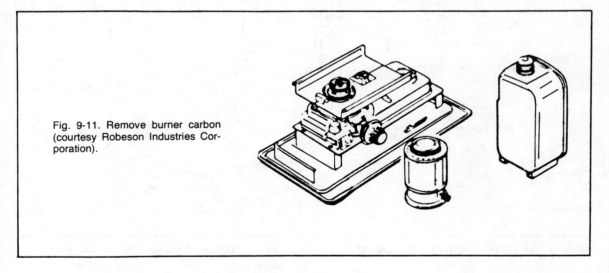

Fig. 9-11. Remove burner carbon (courtesy Robeson Industries Corporation).

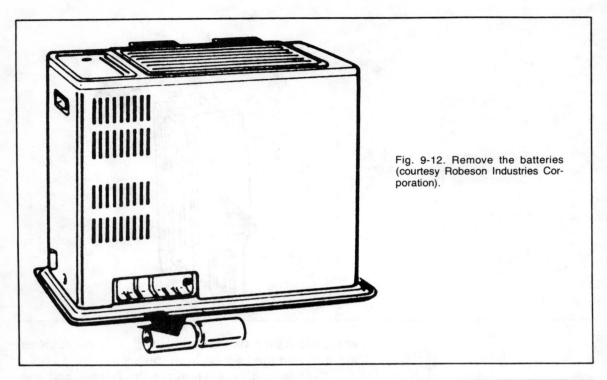

Fig. 9-12. Remove the batteries (courtesy Robeson Industries Corporation).

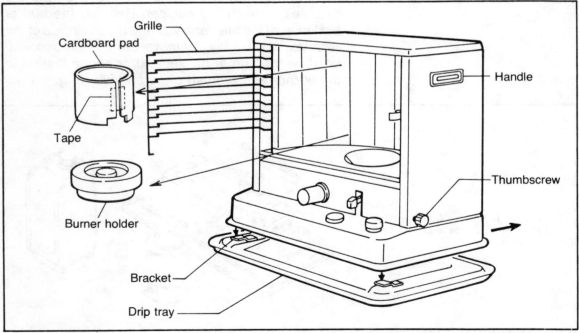

Fig. 9-13. Use original packing material to repack your heater for storage (courtesy Robeson Industries Corporation).

184

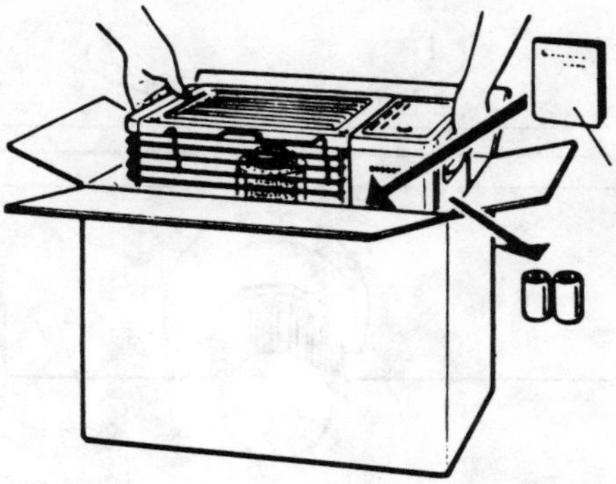

Fig. 9-14. Make sure that the operation manual is included in the storage box (courtesy Robeson Industries Corporation).

Operation manual

After the heater has been thoroughly cleaned, reassemble the unit so the gaps between the burner basket and wick guide cylinder are equal in circumference.

The batteries should be taken out from the battery case because they may leak and corrode the kerosene heater (Fig. 9-12).

Store the kerosene heater with the extinguisher device deactivated and the wick lowered in the heater. This will minimize damage to the unit if it is accidentally jarred while in storage.

Place the kerosene heater in the original packing box as shown in Figs. 9-13 and 9-14. Include the owner's manual. Remove the batteries. Store your portable kerosene heater in a cool, well-ventilated place where it will not be damaged.

Chapter 10

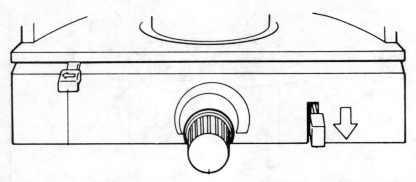

Kerosene Heater Safety Standards

A PORTABLE KEROSENE HEATER IS DEFINED BY the National Fire Protection Association as "an unvented, self-contained, self-supporting heater, with integral reservoir, designed to be carried from one location to another." In recent years, the use of portable kerosene heaters by homeowners has increased dramatically—primarily as a result of their availability, their fuel efficiency, and their advantage in reducing central heating costs by supplemental heating of individual rooms.

Portable kerosene heaters were a frequent cause of loss of life and injury by fire and asphyxiation during the 1960s. This experience prompted many states and municipalities to adopt ordinances prohibiting their use. By the late 1970s, however, models with new safety features began appearing in the marketplace. Many states and municipalities that once prohibited these appliances repealed such restrictions. Conversely, many other states and municipalities with no previous laws on such heaters banned or imposed restrictions on use of the heaters. This disparity in laws has caused much confusion for the consumer and others.

This report, from the Engineering and Safety Service of the American Insurance Association, discusses the hazards associated with portable kerosene heat-

AMERICAN INSURANCE ASSOCIATION REPORT

ers and general precautions that should be followed. The material is from "Portable Kerosene Heaters" and is presented courtesy of the American Insurance Association.

Burner Safety Controls

Portable kerosene heaters should be built to prevent the accidental discharge of fuel in case of ignition failure, flame extinguishment or tip-over. Many older heaters had no burner safety controls and allowed fuel to flood the burner if the unit was not placed exactly level, if no flame was present, or if the heater was tipped over. Most recently manufactured models have their sole fuel supply in a sump or tank located under the burner; fuel reaches the flame via capillary action through a wick. Immediately above the wick is a shutter or a pendulum-actuated latch that closes when the heater is impacted or tipped. It snuffs the flame out and helps keep fuel spillage to a minimum.

Some heaters have what is known as a *barometric feed* of the fuel supply. This method utilizes a removable tank and a sump under the wick. The tank is equipped with a spring-loaded valve. When the tank is in the normal operating position in the heater, it is in an inverted position standing in an open top bowl, which is shaped to receive the tank. The spring-loaded valve is held open so kerosene can bubble out of it whenever the kerosene level in the bowl drops low enough to allow a bubble of air to enter through the valve. The proper liquid level in the bowl thus is maintained. The bowl feeds kerosene to the sump under the burner. The proper liquid level thus is maintained in the sump, from whence the wick absorbs the kerosene, carrying it through capillary action to the flame.

Listed Appliances

If portable kerosene heaters are legal in the state or municipality in question, only those listed by a nationally recognized testing laboratory such as Underwriters Laboratories should be used. Underwriters Laboratories publishes in their *Gas and Oil Equipment Directory* a list of companies that can supply portable kerosene heaters in compliance with their standards. If you are considering the purchase of a portable kerosene heater, look for the listing mark on the appliance.

Underwriters Laboratories listed portable kerosene heaters constructed today contain many safety features to reduce the chance of a fire, burns to people, and the potential asphyxiation problem associated with past heaters. The safety features that UL-listed units must have include:

- A drip tray at the base of the appliance to trap any spills of kerosene.
- A low center of gravity for stability and to reduce the chance of tipping.
- A tip-over sensor that will minimize fuel spillage and will extinguish the burner flame when a unit is tipped or jarred in any direction.
- A protective guard over those parts capable of causing serious burns.
- A fuel tank with a capacity of not more than 2 gallons (7.57 liters) and constructed of materials such that it will not be damaged by being dropped.
- A durable fuel gauge or liquid level indicator to assist in preventing overfilling and unnecessary refilling that will not crack or break if dropped.
- A radiation shield or baffle to prevent excessive temperature transmission to nearby combustibles including those under the appliance.
- A windshield to prevent wind currents or drafts from blowing flames outside the bounds of the appliance.
- Construction such that it will not emit a hazardous concentration of carbon monoxide or carbon dioxide to the living space when the heater is used in accordance with the manufacturer's instructions.

- Controls that are easy for the consumer to operate.
- Instructions that are easy for the consumer to read and understand.
- Follow-up inspections by UL personnel at the factory where these units are manufactured to ensure that the units continue to meet UL requirements.

Ventilation

Portable kerosene heaters are not required to be connected to a chimney, nor are they designed to draw their required oxygen for combustion from outside the building. These heaters draw their required oxygen from the interior of the building in which they are located and return their products of combustion back to the living space. In a home of typical construction—that is, one which is not of unusually tight construction due to heavy insulation and tight seals against air infiltration—an adequate supply of air for combusion and ventilation is provided through infiltration. If the heater is used in small rooms or spaces with poor air infiltration rates or in a well-insulated dwelling, ventilation requirements may become critical. If a heater is used in an area where normal infiltration rates are low, the oxygen in the air will be gradually reduced, resulting in incomplete combustion of the fuel and the liberation of increased amounts of carbon monoxide—which is odorless, colorless, and deadly to occupants if exposure is continuous and at a high enough level.

The manufacturer's literature should be checked, and the dealer consulted, when matching the appropriate size unit for the area intended to be heated. To reduce the risk of asphyxiation, the heater should not exceed a rating of 25,000 Btu (6,300 kilogram-calories) per hour. If the heater is used in a small room where less than 200 cubic feet (5.7 m³) of air space is provided for each 1000 Btu (252 kilogram-calories) per hour of heater rating (considering the maximum burner adjustment), the door(s) to adjacent room(s) should be kept open, or a window to the outside should be opened at least 1 inch (25.4 mm) to guard against carbon monoxide buildup.

Many safety experts believe portable kerosene heaters should not be used in bedrooms or other rooms used for sleeping purposes. These units may be hazardous in bedrooms because sleeping occupants will not be able to observe any abnormal operation of the heater. Because doors of bedrooms should be kept closed for fire safety reasons, the room ventilation requirements for sleeping occupants and for proper operation of the heater may be inadequate. Also, it is unlikely that the practice of partially opening a window in a bedroom and allowing cold air to enter will be followed. The objective of providing the heater is to keep the room warm. People likely will not purchase smaller units for use in their bedrooms. They will probably purchase a unit for the living room or family room and move it to the bedroom—the result being a unit too large for the space intended to be heated.

Kerosene Facts

Only kerosene should be used for fuel in a portable kerosene heater. Fuels such as No. 2 heating oil or machine oil will interfere with proper operation and produce excessive amounts of toxic combustion products that could be lethal. With gasoline, including unleaded or white gasoline and camp fuel, users run the risk of an explosion. Some manufacturers state in their operating instructions that Jet A fuel may be used as a substitute when kerosene is not readily available. Jet A formulations often change seasonally due to the addition of deicers and may also have a high sulfur content. For these reasons, Jet A fuel is not recommended for use in these heaters.

Kerosene is a highly refined petroleum derivative. The grade of kerosene considered safe for use in an unvented heater is grade No. 1-K, a low-sulfur (0.04 percent or less, by weight) kerosene that meets American Society for Testing and Materials (ASTM) D-3699-78, "Standard Specification for Kerosene."

The primary reason for using a grade of kerosene having a limited sulfur content are: to reduce the generation of excessive amounts of sulfur dioxide (a toxic, irritant gas), which at concentrations below lethal limits may be dangerous to the health of persons with asthma or other respiratory problems; and to prolong wick life, avoid caking of the wick, and permit proper burning of the fuel. Only grade No. 2-K (0.05 percent to 3.0 percent sulfur content, by weight) may be available in many regions of the country. If such is the case, the heater should not be purchased.

Kerosene should be clear like water; it should not be yellow or colored. Yellow or colored kerosene indicates inferior or contaminated fuel and is not considered suitable for use in unvented heaters, because it can decrease the wick life, causing smoking and odor problems and increasing risk of asphyxiation. To help assure that kerosene maintains its quality and does not degrade or change chemically, kerosene should only be purchased from a reputable dealer and should not be stored by the user for more than one heating season.

Kerosene Storage

Kerosene is considered a class II combustible liquid, because its flash point is above 100° F. (37.0° C.) and below 140° F. (60° C.). Although a class II combustible liquid must be atomized or wicked to ignite at room temperature, precautions are essential during storage to reduce the chance of a fire or explosion. Kerosene should be stored in approved, metal or plastic containers identified as suitable for storing petroleum products. These containers should be capable of holding no more than 5 gallons (18.9 liters) and be permanently and clearly marked KEROSENE to prevent accidental use of the incorrect fuel. The total amount of kerosene stored should be limited to those quantities permitted by local fire prevention codes. The local fire prevention code should be consulted for provisions concerning the storage locations—both indoors and outdoors. Keep kerosene away from small children.

Refueling Procedures

Refueling a portable kerosene heater can be a dangerous procedure. *A heater should never be refueled while it is still operating or while it is still hot, and the user should never smoke during refueling.* Instructions provided with units constructed with the barometric feed principle may state that the tank can be removed for refilling while the heater is operating (the flame will continue for several minutes on the kerosene in the sump). There is little doubt that many users of the barometric feed type will be replacing the filled tank while the heater continues to operate. This is an undesirable practice, because improper alignment, "hanging up," or any other condition that keeps the unvented tank from being properly positioned can cause fuel overflow. If the heater is operating, the overflow fuel can ignite immediately. If the heater is cool, the overflow fuel can be observed and wiped up before the heater is relighted.

Dispensing kerosene into a portable kerosene heater fuel tank requires that you take care not to spill the fuel or overfill the tank. A siphon pump should be used to keep spillage to a minimum during refueling. The heater or tank should be taken outdoors where spillage will not be a hazard. The pump should contain a filter or strainer so that any impurities in the kerosene are not transferred to the heater. Because cold kerosene will expand as it warms to room temperature, possibly allowing kerosene to overflow if the tank does not have sufficient air space to accommodate expansion, it is also recommended that a tank be filled to more than 90 percent of its capacity. The cap on the tank should always be tightly secured after the refueling procedure.

Placement

Because kerosene heaters are portable, there is danger of fire if the unit is placed too close to combustibles or in normal paths of travel where someone might accidentally trip or fall on it or knock it over. The heaters should never be carried while lighted. They should not be used where flammable vapors or gases may be present, such as in a garage or in the presence of flammable solvents, aerosol sprays, lacquers, etc. The heater should be on a level floor or platform. Some units are designed with their own level indicators. Due to the high surface temperatures on many of these heaters, keep them away from children, clothing, furniture, draperies, curtains, and pets.

Operation

To ensure the safe operation of portable kerosene heaters, you must follow closely the manufacturer's operating instructions. Also, consumers intending to purchase a heater should always have the dealer demonstrate the lighting and operating procedures. While some heaters have a battery-operated, push-button lighting device, others require manual ignition using matches. Details such as trimming the wick of the burner evenly and raising the wick to the correct height are important to prevent smoking or too high a flame (Figs. 10-1 and 10-2).

After lighting a heater, you should always wait for the kerosene to saturate the wick and adjust to an even flame. Watch for smoking or abnormal flame extension for at least another 15 minutes. Operators of portable kerosene heaters sometimes make the mistake of leaving them unattended after lighting. Flame height may intensify over a period due to the vapors from the kerosene increasing as a result of a rise in the temperature of the burner. Even though directions for these appliances often indicate a starting position for the controlling device and another position for operation, it is not a good practice to light an appliance, turn the control to the operating position, and then leave it without further attention.

Maintenance

If you intend to purchase a portable kerosene heater, look for a unit that can be serviced easily and allows you to perform simple maintenance procedures such as wick replacement. These procedures should be demonstrated by the dealer before the

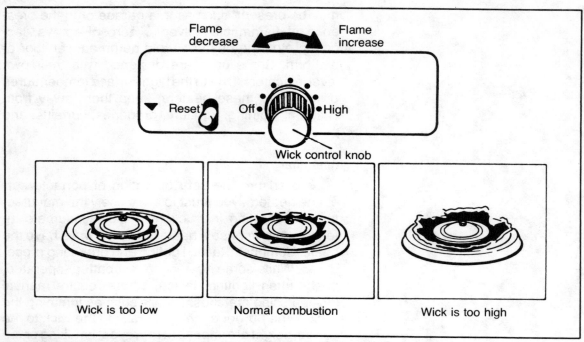

Fig. 10-1. Setting the wick for safety (courtesy Robeson Industries Corporation).

heater is purchased. The safe use of these units also depends on proper assembly (when required) by you, the purchaser. Follow the manufacturer's maintenance instructions. The instruction booklet should always be kept available for easy reference. Worn or damaged parts should be replaced promptly, preferably by the manufacturer or authorized representative. When heaters are to be stored over summer months, all the fuel should be removed from the tank. These units should then be cleaned according to the manufacturer's instructions and subsequently stored in a cool, dry area.

Secondhand Units

Used or secondhand heaters will become more important in the marketplace. Anyone intending to purchase a secondhand heater must look critically at such a unit to make sure it is safe from defects or damage from use by a previous owner. Aside from the recommendations made here, you must be sure that no major alterations have been made to the unit by the previous owner, and that the heater's safety devices still function properly. The manufacturer's instructions should always be transferred along with the unit.

Smoke Detectors and Fire Extinguishers

Because of the inherent danger of fire, smoke detectors should be located as recommended by the manufacturer. At least one multipurpose dry chemical type fire extinguisher should be located in the room where the heater is operating, preferably near the entrance to the room. The operator of the heater should know how to use the fire extinguisher and should move the fire extinguisher along with the heater if the heater is moved. The fire extinguisher should always be available during the refueling procedure.

Summary

Laws that pertain to the use of portable kerosene heaters vary from one locality to another. Although many states and municipalities prohibit the use of such heaters altogether, others may require that they be listed by a nationally recognized testing laborato-

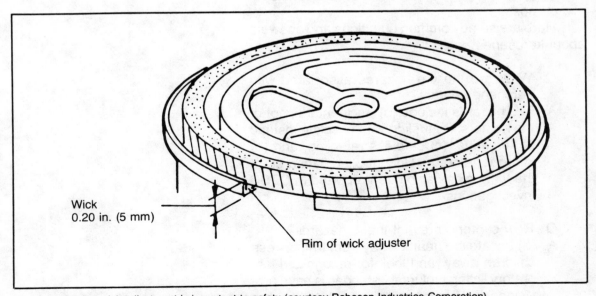

Wick
0.20 in. (5 mm)

Rim of wick adjuster

Fig. 10-2. Correct wick adjustment is important to safety (courtesy Robeson Industries Corporation).

ry or may restrict their use to one- and two-family dwellings. Some have no laws at all covering their use. Before any portable kerosene heater is purchased or used, the local authority having jurisdiction or fire department should be consulted.

This information is from "Product Safety Fact Sheet No. 97: Kerosene Heaters, December 1982" courtesy of the U.S. Consumer Product Safety Commission, Washington, DC 20207.

The use of kerosene portable heaters has expanded significantly over the past few years; an estimated 3 million heaters were sold in 1981 and 10 to 12 million by the end of 1982.

The major potential hazards associated with kerosene heaters are risk of fire and exposure to possibly harmful levels of carbon monoxide, sulfur dioxide, and nitrogen oxides. In addition, the use of gasoline or kerosene contaminated with gasoline or some other substance more flammable than kerosene creates a serious fire hazard. If fuel containers become confused, consumers may unknowingly use gasoline. Consumers are advised to purchase and store kerosene in metal containers clearly marked KEROSENE and never to use red gasoline cans for this purpose.

Here are some common questions and answers about kerosene heaters:

Q. What are the fire hazards associated with kerosene heaters?
A. Fire hazards include (a) accidental upset of the heaters, (b) fueling with gasoline instead of kerosene, (c) careless refueling, and (d) improper storage. Uncontrolled flaring up of the flames has also been reported in CPSC investigations.

Q. How can one prevent these hazards?
A. (a) Locate the heater away from traffic, keep children away, and look for recognized laboratory listing (voluntary standards have provisions for heater stability.) (b) *Never* use

KEROSENE HEATERS PRODUCT SAFETY FACT SHEET FROM THE U.S. CONSUMER PRODUCT SAFETY COMMISSION

gasoline and be aware of its potential for uncontrolled fire. (c) Follow manufacturer's instructions, fill outside, wipe up spills, do not use old or contaminated kerosene, and never fill a hot heater. (d) Use proper containers clearly marked KEROSENE, preferably metal (not red gasoline cans because of the possibility of confusion), and store them in outside living areas. If flare-up occurs and the flame will not extinguish, leave the area and call the fire department.

Q. What if a heater leaks or the flame does not extinguish?
A. Such heaters may be defective and should be returned to the dealer. The CPSC is investigating heaters to determine whether a substantial product defect exists and would like to hear of any incidents.

Q. What are the health effects produced by emitted pollutants?
A. The health effects could vary widely, depending upon which pollutants are produced and in what concentrations. The CPSC is not aware of studies of health effects specifically associated with the use of these heaters. Health effects known to be associated with the pollutants that can be emitted—carbon monoxide, nitrogen oxides, sulfur dioxide, and formaldehyde—are concentration dependent. Irritation of the nose, eyes, and throat, headaches, drowsiness, difficulty in breathing, and impaired lung functioning are associated with exposure to these pollutants. The CPSC is also concerned about the long-term effects of carbon monoxide exposure, especially on angina patients; the ability of nitrogen dioxide and other combustion products to cause an increase in respiratory disease in children; and the potential carcinogenicity of formaldehyde. The staff is

evaluating the potential for adverse health effects from these heaters.

Q. How can one minimize any potential effects?

A. Use the correct fuel wick. Do not use an oversized heater for the room. Provide sufficient ventilation. Do not run the heater overnight or when asleep. Discontinue use if discomfort is experienced. The CPSC recommends that consumers use kerosene heaters sparingly; use adequate ventilation; consider placing a radiant heater inside a fireplace if available; and not use home heating oil, gasoline, or any other fuel except that recommended by the manufacturer.

Q. Why doesn't CPSC ban the heaters now?

A. The CPSC does not have information showing that they should be banned. The priority efforts now underway will assist consumers and industry in defining the potential hazards involved—thermal burns, fire, indoor air pollution—and determine the need, if any, to reduce such hazards.

The U.S. Consumer Product Safety Commission offers these basic suggestions to consumers.

Selection

- Before purchasing a heater, make sure that local building and fire codes permit the use of kerosene heaters in residences.
- Buy a heater that has been labeled and listed as tested by a nationally recognized testing organization.
- Select a heater appropriately sized to the room you want to heat. Do not purchase heaters too large for the room.
- Select a heater with a safety shutoff device that automatically snuffs out the flame if the heater is tipped over (Fig. 10-3). Some heaters also have a push-button lighting device that eliminates the need for matches.

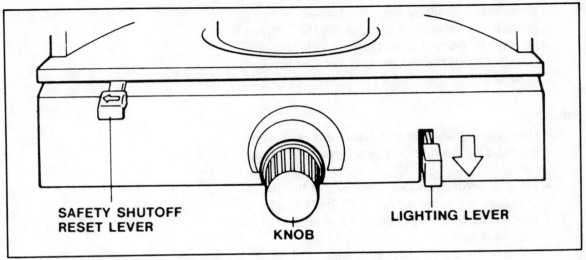

SAFETY SHUTOFF
RESET LEVER

KNOB

LIGHTING LEVER

Fig. 10-3. Safety shutoff reset lever (courtesy Robeson Industries Corporation).

Use

- Kerosene heaters are for supplemental heat only. Do not operate them unattended.
- Read and follow the manufacturers' instructions carefully.
- Place heaters on a level, protected surface where they are not likely to be bumped or knocked over.
- Insure adequate ventilation even if you have to open a window slightly. Leave the door to the rest of the house open. Never use any unvented heater in a closed room.
- Turn off the heater before going to sleep.
- At high levels, the pollutants produced may cause adverse health effects. If dizziness, drowsiness, chest pain, fainting, or respiratory irritation occurs, shut off the heater and move the victim to fresh air.
- Keep the heater away from combustible walls and furnishings such as furniture, drapes, or easily ignited materials like paper. Keep clothing away from any heater.
- Turn the heater off when leaving the room. Do not leave children or pets in the same room with a heater unless an adult is present.

- Keep children away from the heater.
- In case of a flare-up or uncontrolled fire, do not attempt to carry the burning heater from the house. Leave the house and call the fire department.

Refueling

- Use only water-clear K-1 kerosene as recommended by the manufacturer. Never use gasoline.
- Fill the heater outdoors. Allow the heater to cool before refueling. Do not fill the tank all the way to allow for expansion of fuel and to prevent leakage.
- If there is a spill or leakage, wipe up fuel immediately.

Maintenance

- Inspect the heater frequently for proper adjustment; clean the heater regularly.
- Always follow manufacturer's recommendations for periodic maintenance.

Storage

- Never use a gas can to store kerosene. The kerosene may become contaminated with gasoline that remains in the can, and an uncontrolled fire may result.
- Store kerosene outside the house in a properly labeled container to avoid a mix-up with a gasoline container.
- Store kerosene out of the reach of children. Kerosene may be fatal to children if ingested. Chemcial pneumonitis results from aspiration into the lungs.
- If under 5 gallons, the container used must have safety closures.

Reporting Product Hazards

To report a product hazard or a product-related injury, write to the U.S. Consumer Product Safety

Commission, Washington, DC 20207. In the continental United States, call the toll-free hotline 800-638-CPSC (2772). A teletypewriter for the deaf is available by calling the following numbers: national (including Alaska and Hawaii) 800-638-8270, Maryland residents only 800-638-8104. During nonworking hours, messages are recorded and answered on the following day.

You may also contact regional offices of the U.S. Consumer Product Safety Commission listed in metropolitan phone books.

The U.S. Consumer Product Safety Commission (CPSC) is an independent regulatory agency charged with reducing unreasonable risks of injury associated with consumer products. The CPSC is headed by five commissioners appointed by the President with the advice and consent of the Senate.

SUMMARY DESCRIPTION OF TESTING REQUIREMENTS FOR PORTABLE KEROSENE HEATERS: UNDERWRITERS LABORATORIES SUBJECT NO. 647

This testing information is important for the safe operation of portable kerosene heaters. It is provided courtesy of Kero-Sun, Inc., Main Street, Kent, CT 06757.

Fuel Tanks

The capacity of the fuel tank must not exceed 2 gallons.

Temperature Measurement and Lighting Tests

A series of tests is conducted in which the surface temperature of heater components and the test enclosure are measured. In addition, tests are conducted to determine the effect of exposure to combustible materials under various conditions of lighting and operating procedures.

Lighting Test

The heater must be capable of being lighted without flash or puff while still hot from previous operations.

Combustion Test

Units are tested for stable and complete combustion, and heating surfaces are checked for deposits of soot and tar.

Stability

Units are tested with a full fuel reservoir to determine that they can be tilted at least 33 degrees (from the vertical axis) without tipping over.

Tip-Over Test

An operating heater must exhibit its capability to undergo repeated tip-overs onto fabric-covered floors without causing fire.

Tip-Over Switch Endurance Test

Heater extinguishers must successfully complete 6,000 trigger operations without evidence of undue wear.

Aging and Immersion Tests

Tests determine resistance of polymeric materials and any gaskets to deterioration from heat, kerosene, and oxidation.

Marking Plate Adhesion Test

Tests determine durability of the adhesive-secured marking plate under ordinary usage and its resistance to defacement after oven aging or immersion in water or kerosene.

Manufacturing and Production Tests

Manufacturers are required to provide regular production control, inspection, and tests. At least four times a year, UL conducts follow-up service inspections to insure that raw materials and manufactured parts continue to meet original specifications.

Warning Markings

The following cautionary markings (or equivalents) must appear on all portable kerosene heaters:

—CAUTION: Hot while in operation. Do not

touch. Keep children, clothing, and furniture away.

—CAUTION: Risk of asphyxiation. Use this heater only in a well-ventilated area.

—CAUTION: Risk of fire. Do not remove fuel tanks when the heater is operating or when it is hot (applicable only to heaters with removable tanks).

The following additional tests are included in a proposed revision of subject No. 647:

Carbon Monoxide Emissions

When heaters are operating at maximum heat output, carbon monoxide concentration cannot exceed 0.04 percent of an air-free sample of the flue gases. When the heater is operating at the recommended low-fire setting, the carbon monoxide concentration cannot exceed 0.08 percent of an air-free sample of the flue gases.

Normal Fuel Level Test

When operating at maximum fuel input, the heat cannot cause fuel to expand to a point where overflow may occur. When operating, heaters with removable cartridge tanks are tested at maximum fuel input to determine that heat cannot cause fuel to expand and produce subsequent leakage.

Requests for complete testing information should be addressed to: Heating, Air Conditioning, and Refrigeration Department, Underwriters Laboratories, 333 Pfingsten Road, Northbrook, IL 60062.

Glossary

air change—Replacement of all the air in a room over a period.

air filter—A device used to clean dust particles from the air as it is circulated through a furnace heating system. Some filters are disposable. Others are permanent and can be vacuumed or washed.

air supply system—Air is supplied to a portable kerosene heater through the air tube to the inside of the circle of flames. Air is also supplied by the outer and middle tubes to the air outside the circle of flames.

ambient temperature—Refers to the outside temperature.

batts—A type of insulation made from mineral or glass fiber that comes in precut sheets to fit standard stud spaces.

blankets—A type of insulation made from mineral or glass fiber that comes in rolls.

Btu—British thermal unit; a unit for measuring a quantity of heat. One Btu is the amount of heat required to raise the temperature of 1 pound of water 1 degree Fahrenheit—approximately the heat released by totally burning one wooden kitchen match.

Btu/hr.—British thermal units per hour refers to heat generation. Heating and cooling systems are rated in Btu/hr.

caulking—A pliable material used to seal up gaps and cracks in structures.

Celsius—Temperature scale that registers the freezing point of water at 0° C. and its boiling point at 100° C.

condensation—When water vapor comes in contact with a cold surface, the vapor turns to droplets of water.

conduction—Heat transfer through a material in which energy is transmitted from particle to particle without displacement of the particle.

convection—The transmission of heat by the mass movement of heated gases or liquids through contact with other gases, liquids, or solids at higher or lower temperatures.

convection heater—A portable kerosene heater that transfers heat to the air through the convection method.

degree-day—A unit, based on the temperature difference and time, used in describing how warm or cold a period of time was. This unit is used for estimating a heating system's energy consumption. For any one day whose mean temperature is below 65° F., the degree-day is the difference between 65° F. and the mean for that day.

duct—A pipe or closed conduit made of sheet metal

or other suitable material used for conducting air to and from an air-handling unit.

exhaust method—The method by which exhaust gases are expelled by a portable kerosene heater. The radiant heater exhaust method expels exhaust gases inside the room naturally using the combustion tube draft. The convection heater expels exhaust gases inside the room using the draft of the burner assembly.

Fahrenheit—Temperature scale that registers the freezing point of water at 32° F. and the boiling point of water at 212° F.

flue—A pipe, tube, or channel through which hot air, gas, steam, or smoke may pass as in a boiler or a chimney.

fuel supply system—The kerosene is supplied to the top of the wick using the capillary tube effect of the wick from the fixed kerosene surface formed by the tank(s).

furnace—A unit used to heat air.

glazing, double— A single glazed window with an additional piece of glass installed on the sash to provide an air space between them. The second glass can either be fixed or movable and can be installed on the inside or outside of the sash. Double glazing differs from insulating glass in that there is no permanent seal around the panes to create a true dead air space.

glazing, triple—A sash glazed with three panes of glass, enclosing two separate air spaces.

heat conductor—Any material that allows heat to pass through it quickly.

heat exfiltration—When the outside temperature is colder than the inside of a house, heated air will leak outside through cracks and gaps in the structure.

heat gain—The heat that flows into or is released within a structure due to solar radiation, waste heat from motors, lighting, or people.

heat infiltration—When the outside temperature is warmer than the inside of a house, warm air will leak inside through cracks and gaps in the structure.

heat loss—The condition in houses caused by heat's physical property of always moving to lower temperatures by leaking out of the house.

ignition—When the kerosene heater's wick is ignited or lit by either a match or an ignition coil and battery.

insulation—A material having a relatively high resistance to heat flow and used principally to retard the loss of heat from a home.

kilowatt—1000 watts of electricity.

kilowatt-hour—1000 watts of electricity used for an hour's time. The basic unit of measurement used by electric meters.

loose fill—An insulating material suitable for blowing through a machine into attics or walls. Some loose fill may be poured.

pilot light—A continuously burning flame used to turn on gas furnaces, ovens, and ranges.

radiant heater—A kerosene heater that emits heat through radiation.

radiation—The transmission of energy by means of electromagnetic waves of generally long wavelength. Radiant energy of any wavelength may become thermal energy when absorbed and result in an increase in the temperature of the absorbing body.

relative humidity—This is a measure, in percentage form, indicating the amount of water vapor in the air compared to the amount the air could hold if it was completely saturated with moisture.

R value—A number used to describe the resistance of a material to the flow of heat. The larger the R value of material, the higher the insulation value.

safety mechanism—The automatic extinguishing device on a portable kerosene heater.

therm—A quantity of heat equal to 100,000 Btu.

thermal envelope—A term generally used when describing the walls, windows, doors, ceilings, and floors around heated areas of a structure. By improving the thermal envelope, you can increase the structure's comfort and reduce energy requirements needed for heating and cooling.

thermostat—An instrument that responds to changes in temperature and automatically directs temperature control. In a home heating system it signals the baseboards or furnace when heat is required to maintain present temperature settings. It can be used with a portable kerosene heater by sensing how much heat is required above that developed by the supplementary heater to maintain the controlled temperature.

vapor barrier—A material used to retard the transfer of water vapor from one area to another. It is used to retard water vapor transfer through walls and through the floor from the ground below the structure.

vent—Anything that allows air to flow in and out of a space.

ventilation—Bringing air into a space, such as a house, with some controlled mechanical or natural system.

watt—A unit of power, especially electric power.

weather stripping—Metal, plastic, or foam strips designed to seal between windows and door frames to prevent air leakage.

wick—Bundles of thin fibers and capillary tubes designed to draw kerosene up from the tank into the combustion area of a kerosene heater. The top of a kerosene heater wick is typically of fiberglass strands, and the bottom is of cotton.

Index